About the series

100 Mental Maths Starters is a series of six photocopiable teacher's resource books, one for each of Years 1–6. Each book offers 100 mental maths activities, each designed to last between 5 and 10 minutes. These activities are ideal to start your daily dedicated maths lessons if you are following the National Numeracy Strategy. Each year-specific book provides mental activities for maths within the guidelines of the NNS *Framework for Teaching Mathematics*. The activities can also be used effectively to meet the needs of Primary 1–7 classes in Scottish schools, or classes in other schools functioning outside the boundaries of the National Numeracy Strategy.

This series provides suitable questions to deliver the 'Oral and Mental Starters' outlined in the lesson plans in the companion series from Scholastic, *100 Maths Lessons and more*. Reference grids are provided (see pages 4–5) to indicate the lesson and page numbers of the associated lesson plans in the relevant *100 Maths Lessons and more* book. However, the series is also wholly appropriate for independent use alongside any maths scheme of work. The index at the back of each book makes it easy to choose a suitable Starter activity for any maths lesson.

Each book provides support for teachers through three terms of mental maths, developing and practising skills that will have been introduced, explained and explored in your main maths lesson time. Few resources are needed, and the questions for each activity are provided in full. The books are complete with answers, ready for you to pick up and use. In addition, all the activities in the book can be photocopied and the answers cut off to leave activity cards that pupils can work from individually. Alternatively, the activity cards can be used by pairs or small groups, with one child asking questions and the other(s) trying to answer.

The activities are suitable for use with single- or mixed-ability groups and single-age or mixed-age classes, as much emphasis has been placed on the use of differentiated and open-ended questions. Differentiated questions ensure that all the children can be included in each lesson and have the chance to succeed; suitable questions can be directed at chosen individuals, almost guaranteeing success and thus increased confidence.

Learning mental maths

The mental maths starters in this book provide a structured programme with a balanced progression. They provide regular opportunities for all children to learn, practise and remember number facts. Completed in order, the activities in this book provide the framework of a scheme of work for mental maths practice in Year 3. Some essential photocopiable resource pages are also included (see pages 76–9). To cover the whole year, you will need to add some repeats and/or variations of the activities for consolidation (for which purpose some alternative sets of questions are provided). This book does not provide the groundwork concept teaching for each new skill: that is covered in detail as the focus of appropriately timed main teaching activities, in *100 Maths Lessons and more: Year 3* by Sue Gardner and Ian Gardner.

Each activity in this book has one or more learning objectives based on the 'Teaching Programme: Year 3' in the NNS *Framework*. Key objectives are highlighted in bold. Teacher instructions are provided, stating the particular strategies being developed or encouraged. Discussion of the children's methods is emphasised, since this is essential: it will help the children to develop mathematical language skills; to appreciate that no single method is necessarily 'correct', and that a flexible repertoire of approaches is useful; and to improve their overall confidence as they come to realise that all responses have value. Strategies are encouraged that will enable children to progress from known to unknown number facts, thus developing their ability to select and use methods of mental calculation.

As adults, we probably do maths 'in our heads' more often than we use written methods. Almost without thinking about it, we apply flexible strategies that we developed as children. By following the activities in this series, children will learn to explain their thought processes and techniques – which, in turn, will help them to clarify their thinking and select appropriate methods to use in different contexts.

About this book

This book is aimed at developing the mental and oral skills of pupils in Year 3/Primary 3–4. It builds on the work of earlier years, covered in the books for Years 1 and 2/Primary 1–3 in this series. The activities in this book provide a sound basis for work in Year 4/Primary 5 and subsequent years, covered by the later books.

In this book, emphasis is placed on strategies for addition and subtraction (especially with numbers up to 50); adding from the larger number; counting up for a small difference; partitioning; and using doubles and 'near' doubles. Repeated opportunities for counting in 2s, 5s and 10s will help to develop the children's understanding of patterns in times tables. The relationships between different times tables are also used to support learning of new tables. Some lessons are based on simple fractions and their equivalents, telling the time, and exploring number patterns. Games are included in each term's work to help provide variety and generate enthusiasm for numbers. Open-ended questions are used to challenge the children and extend their thinking.

This book is, first and foremost, a resource for practising teachers. Comments and suggestions from teachers using the book will thus be very welcome, and may be incorporated into future editions.

	100 Mental Maths Starters		100 Maths Lessons		
Starter activity	Activity	Term	Unit	Lesson	Page
1	reading numbers: place value	1	1	1	19
2	addition and subtraction: making 10	1	1	2	20
3	addition: bridging 10	1	1	3	21
4	addition: counting on from the larger number	1	2–3	1	25
5	adding more than two numbers: making 10	1	2–3	2	26
6	subtraction by counting up	1	2–3	3	27
7	doubles of numbers to 10	1	2–3	4	27
8	doubles of numbers to 13	1	2–3	5	28
9	addition: in any order	1	2–3	7	29
10	subtraction: complementary addition	1	2–3	8	30
11	addition and subtraction facts to 20	1	4–6	1	34
12	addition facts making 20	1	4–6	3	35
13	addition facts to 20	1	4–6	6	37
14	subtraction facts to 20	1	4–6	7	38
15	doubles of numbers to 15 and corresponding halves	1	4–6	8	38
16	subtraction as inverse of addition	1	4–6	9	39
17	money: using coins to 50p	1	4–6	11	40
18	money: cost of more than one item	1	4–6	12	40
19	money: word problems	1	4–6	13	40
20	addition grid: odd and even numbers	1	8	2	50
21	addition grid: subtraction as inverse of addition	1	8	4	51
22	money: using £1 coin and silver coins	1	9–10	1	56
23	addition: facts making 50	1	9–10	2	57
24	money: using all coins to £1	1	9–10	4	58
25	2 times table	1	9–10	8	60
26	4 times table	1	9–10	9	61
27	magic square: number puzzle	1	11	1	66
28	5 times table	1	11	3	68
29	multiples of 2, 5 and 10	1	11	5	69
30	doubles of numbers to 20	1	12	1	71
31	subtraction as inverse of addition	1	12	2	72
32	doubles and halves	1	12	3	73
33	5 times table	1	13	1	75
34	4 times table	1	13	2	76
35	3 times table	1	13	3	77
36	2, 5 and 10 times tables	1	13	4	78
37	addition: facts to 20	2	1	1	87
38	subtraction: facts to 20	2	1	2	88
39	addition: bridging a multiple of 10	2	1	3	89
40	subtraction as inverse of addition	2	2–3	1	92
41	doubles of numbers to 10	2	2–3	2	92
42	subtraction and related vocabulary	2	2–3	4	93
43	word problems	2	2–3	6	95
44	doubling and halving	2	2–3	9	97
45	doubling and halving	2	2–3	10	97
46	finding a quarter; relating a quarter to a half	2	4–6	2	104
47	equivalent fractions: halves and quarters	2	4–6	3	104
48	equivalent fractions: halves, quarters, eighths	Term	4–6	Lesson	105
49	equivalent fractions: halves, quarters, eighths	2	4–6	5	106
50	2 and 5 times tables	2	4–6	6	107

100 Mental Maths Starters		100 Maths Lessons			
Starter activity	Activity	Term	Unit	Lesson	Page
51	2 times table	2	4–6	9	108
52	5 times table	2	4–6	10	109
53	10 times table; pairs of multiples of 100 totalling 1000	2	4–6	11	110
54	2 and 5 times tables	2	4–6	13	111
55	mixed operation puzzle	2	8	1	119
56	addition: explain methods and reasoning	2	8	2	120
57	subtraction: explain methods and reasoning	2	8	4	121
58	addition: multiples and near multiples of 10	2	9–10	3	126
59	addition: near multiples of 10	2	9–10	4	127
60	1–100 grid: addition strategies	2	9–10	6	128
61	addition: re-ordering	2	9–10	8	129
62	subtraction: multiples and near multiples of 10	2	9–10	10	131
63	5 times table	2	11	1	134
64	adding 1, 10 or 100	2	11	2	135
65	place value: multiplying by 10	2	11	3	135
66	3 times table	2	11	4	136
67	comparing and ordering numbers	2	12	1	139
68	rounding numbers to 10 and to 100	2	12	5	142
69	counting in threes	3	1	1	149
70	4 times table	3	1	2	150
71	3 and 4 times tables	3	2–3	1	153
72	complementary addition: 1–100 grid	3	2–3	2	154
73	complementary addition: 1–100 grid	33	2–3	5	155
74	doubles of numbers to 10	3	2–3	6	156
75	doubles of numbers to 20	3	2–3	7	157
76	identifying near doubles	3	2–3	8	158
77	identifying near doubles	3	2–3	10	159
78	pairs of multiples of 5 totalling 100	3	4–6	2	163
79	pairs of multiples of 5 totalling 100	3	4–6	3	163
80	addition and subtraction facts to 20	3	4–6	4	164
81	place value: multiplying by 10	3	4–6	6	165
82	place value: multiplying by 100	3	4–6	7	166
83	doubling and halving	3	4–6	9	167
84	division by 2 and 10	3	4–6	12	169
85	division by 5	3	4–6	13	170
86	halving twice to find a quarter	3	8	1	177
87	halving twice	3	8	2	178
88	multiplying by 10 and 100	3	8	3	179
89	dividing by 10 and 100	3	8	5	180
90	3 times table	3	9–10	1	183
91	multiplication and division by 3	3	9–10	6	186
92	multiplication and division by 4	3	9–10	8	188
93	addition: in any order	3	11	1	192
94	multiplication: in any order	3	11	3	193
95	reading times to 5 minutes	3	12	1	197
96	showing times to 5 minutes	3	12	4	199
97	showing and reading times to 5 minutes	3	12	5	200
98	multiplication by 10	Term	13	1	201
99	division by 10 and 100	3	13	3	203
100	multiplication and division by 10 and 100	3	13	5	205

HTU chart

Starter activity 1

Resources
An HTU chart (enlarged to at least A3 from photocopiable page 76), a pointer.

Objective
Read whole numbers in figures and say them.

Strategies
● Count along the top row of numbers on the chart, then point to individual numbers for children to read out.
● Combine numbers by pointing, eg 200 then 3, for children to say ('two hundred and three').

● Now include 'tens' numbers from the top row of 'tens' on the chart. These use the sound of the unit figure (eg 40 is 'four-ty').
● Go on to use 'tens' numbers from the second row of 'tens'. These have irregular sounds (eg 50 is 'fifty' not 'five-ty').

1.	400 7	8.	100 4	15.	400 8		
2.	300 9	9.	900 6	16.	300 4		
3.	600 2	10.	200 5	17.	900 1		
4.	800 5	11.	800 3	18.	600 7		
5.	700 8	12.	100 6	19.	800 1		
6.	200 1	13.	500 9	20.	500 5		
7.	500 3	14.	700 2				

21.	40 8	31.	400 60 5	41.	300 50 8
22.	90 1	32.	700 80 2	42.	800 30 3
23.	70 3	33.	500 70 4	43.	500 10 7
24.	60 4	34.	100 90 3	44.	100 20 5
25.	30 6	35.	300 40 6	45.	900 30 1
26.	20 2	36.	200 70 1	46.	600 10 9
27.	80 7	37.	900 60 8	47.	200 50 2
28.	10 5	38.	600 80 9	48.	700 20 4
29.	50 9	39.	800 40 7	49.	400 10 1
30.	70 6	40.	100 90 6	50.	800 30 5

Spider maths

7 + 3

1 + 9

6 + 4

12 – 2

Starter activity 2

Resources
A board or flip chart.

Objective
Understand the operations of addition and subtraction; use the related vocabulary.

Strategies
● Write a target number (eg 10) in the middle of a 'spider' diagram. Ask the class to suggest different ways of making that amount. Individual children can write on the board.
● Start with addition. Remind the children that addition can be done in any order.
● Now let them try using subtraction, using facts that start with a number up to 20.

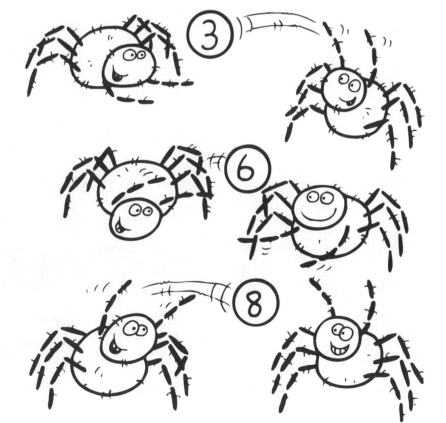

Cross the bridge

1. 18
2. 17
3. 14
4. 13
5. 15
6. 10
7. 12
8. 16
9. 11
10. 19

11. 14
12. 14
13. 12
14. 15
15. 12
16. 14
17. 12
18. 11
19. 11
20. 16
21. 14
22. 13
23. 14
24. 17
25. 11
26. 12
27. 11
28. 15
29. 11
30. 16

Starter activity 3

Resources
A board or flip chart.

Objective
Add a pair of numbers mentally by bridging through 10.

Strategies
● Remind the children that adding on to make 10 and then going into the teens (bridging through 10) is easy.
● Practise adding numbers to 10 to build the children's confidence.

● Demonstrate 6 + 7 on the board, partitioning 7 into 4 + 3 so the 4 can be added first.

1. $10 + 8$
2. $10 + 7$
3. $10 + 4$
4. $10 + 3$
5. $10 + 5$

6. $10 + 0$
7. $10 + 2$
8. $10 + 6$
9. $10 + 1$
10. $10 + 9$

11. $5 + 9$
12. $8 + 6$
13. $7 + 5$
14. $6 + 9$
15. $4 + 8$
16. $9 + 5$
17. $6 + 6$
18. $8 + 3$
19. $3 + 8$
20. $9 + 7$

21. $6 + 8$
22. $4 + 9$
23. $7 + 7$
24. $8 + 9$
25. $4 + 7$
26. $9 + 3$
27. $5 + 6$
28. $8 + 7$
29. $2 + 9$
30. $8 + 8$

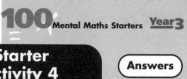
Count on

1. What is 25 more than 4?

2. What is 32 more than 6?

3. What is 20 more than 7?

4. What is 23 more than 5?

5. 8 add 27

6. 7 add 24

7. 9 add 35

8. 6 add 26

9. 7 add 37

10. 3 add 29

11. 8 add 34

12. 9 add 22

13. What is 5 more than 36?

14. What is 3 more than 48?

15. What is 7 more than 39?

16. What is 42 more than 3?

17. What is 57 more than 2?

18. What is 46 more than 5?

19. What is 58 more than 4?

20. What is 44 more than 6?

21. What is 6 more than 38?

22. What is 10 more than 41?

23. What is 9 more than 54?

24. What is 7 more than 33?

25. What is 9 more than 28?

26. What is 7 more than 45?

27. 52 add 8

28. 21 add 9

29. 49 add 9

30. 53 add 6

Starter activity 4

Objectives
Understand the operation of addition and the related vocabulary. Understand that addition can be done in any order.

Strategies
● Ask 'What is 26 more than 3?' to demonstrate that it is easier to put the larger number first and then count on by the smaller number.

Answers

1. 29
2. 38
3. 27
4. 28
5. 35
6. 31
7. 44
8. 32
9. 44
10. 32
11. 42
12. 31
13. 41
14. 51
15. 46
16. 45
17. 59
18. 51
19. 62
20. 50
21. 44
22. 51
23. 63
24. 40
25. 37
26. 52
27. 60
28. 30
29. 58
30. 59

Add three or more numbers

Answers

1. 16
2. 13
3. 14
4. 12
5. 18
6. 15
7. 19
8. 17
9. 12
10. 19
11. 15
12. 13
13. 14
14. 13
15. 12
16. 13
17. 17
18. 15
19. 12
20. 11
21. 20
22. 20
23. 19
24. 16
25. 17

Starter activity 5

Resources
A board or flip chart.

Objective
Understand that more than two numbers can be added together.

Strategies
• Ask for pairs of numbers that total 10.
• Write up $3 + 5 + 7 =$. Ask for the answer and strategies to find it.
• Stress that making 10 (eg $3 + 7$) and changing the order (eg $3 + 5 + 7 = 3 + 7 + 5 = 10 + 5$) can be useful strategies when adding three or more numbers.

1. $2 + 8 + 6$
2. $9 + 1 + 3$
3. $5 + 4 + 5$
4. $7 + 2 + 3$
5. $8 + 6 + 4$
6. $5 + 9 + 1$
7. $9 + 8 + 2$
8. $5 + 5 + 7$
9. $4 + 2 + 6$
10. $2 + 8 + 9$
11. $6 + 5 + 4$
12. $3 + 7 + 3$
13. $4 + 1 + 9$

14. $4 + 6 + 3$
15. $2 + 3 + 7$
16. $3 + 6 + 4$
17. $1 + 9 + 7$
18. $8 + 5 + 2$
19. $3 + 2 + 7$
20. $6 + 4 + 1$
21. $2 + 8 + 3 + 7$
22. $9 + 1 + 6 + 4$
23. $8 + 2 + 3 + 6$
24. $4 + 3 + 6 + 3$
25. $2 + 7 + 5 + 3$

SCHOLASTIC

Count on

1. 27 – 22 =
2. 29 – 25 =
3. 35 – 31 =
4. 38 – 33 =
5. 32 – 29 =
6. 23 – 15 =
7. 24 – 19 =
8. 33 – 28 =
9. 21 – 15 =
10. 42 – 38 =
11. 26 – 18 =
12. 32 – 27 =
13. 42 – 36 =
14. 51 – 47 =
15. 53 – 46 =
16. 47 – 39 =
17. 42 – 35 =
18. 53 – 47 =
19. 54 – 48 =
20. 35 – 29 =

Starter activity 6

Resources
A board or flip chart.

Objective
Find a small difference by counting up.

Strategies
● Write up 31 – 28 = .
Ask for the answer and strategies to solve it.
● Stress that 'counting on' can be useful for subtraction of numbers that are close together (eg 28, 29, 30, 31, so 31 – 28 = 3).

Answers
1. 5
2. 4
3. 4
4. 5
5. 3
6. 8
7. 5
8. 5
9. 6
10. 4
11. 8
12. 5
13. 6
14. 4
15. 7
16. 8
17. 7
18. 6
19. 6
20. 6

Double up

1. 2 times 4
2. double 5
3. twice 11
4. add 1 to itself
5. twice 7
6. double 10
7. twice 13
8. 2 times 8
9. add 11 to itself
10. add 6 to itself
11. twice 9
12. add 0 to 0
13. double 7
14. add 13 to itself
15. 2 times 3
16. twice 6
17. add 14 to 14
18. double 12
19. double 2
20. add 8 to itself

Starter activity 7

Objective
Derive quickly doubles of numbers to at least 10.

Strategies
● Make sure the children are familiar with various words and phrases that mean 'doubling': twice, 2 times, double, add a number to itself.

Answers
1. 8
2. 10
3. 22
4. 2
5. 14
6. 20
7. 26
8. 16
9. 22
10. 12
11. 18
12. 0
13. 14
14. 26
15. 6
16. 12
17. 28
18. 24
19. 4
20. 16

SCHOLASTIC

11

More doubles

Starter activity 8

Resources
The numbers 0–10 (inclusive) written on a board or OHT, well spaced out and at child height.

Objective
Derive quickly doubles between 10 and 20.

Strategies
● Ask for a volunteer to write the double of any number (0–10) beneath that number on the board or OHT. Continue until they are all done (each volunteer writes one answer).
● Ask quickfire whole-class questions while the class can see the answers.

● Gradually erase the better-known doubles while repeating questions 1–10.
● Erase all the doubles and ask questions 11–30, with the children raising their hands to answer individually.

1. 6	
2. 16	
3. 10	
4. 2	
5. 14	
6. 0	
7. 8	
8. 18	
9. 4	
10. 12	
11. 10	
12. 6	
13. 16	
14. 2	
15. 14	
16. 20	
17. 12	
18. 18	
19. 4	
20. 0	
21. 8	
22. 18	
23. 24	
24. 10	
25. 28	
26. 22	
27. 0	
28. 26	
29. 14	
30. 24	

1. double 3
2. double 8
3. double 5
4. double 1
5. double 7

6. double 0
7. double 4
8. double 9
9. double 2
10. double 6

11. 2 times 5
12. add 3 to itself
13. twice 8
14. double 1
15. add 7 to 7
16. 2 times 10
17. double 6
18. add 9 to itself
19. twice 2
20. double 0

21. add 4 to itself
22. twice 9 is
23. double 12
24. add 5 to itself
25. 2 times 14
26. twice 11
27. add 0 to itself
28. double 13
29. 2 times 7
30. add 12 to itself

Number strings

1. 5 + 5 + 3

2. 8 + 1 + 2

3. 7 + 3 + 4

4. 3 + 1 + 9

5. 6 + 1 + 4

6. 5 + 2 + 5

7. 4 + 6 + 9

8. 7 + 2 + 8

9. 3 + 7 + 8

10. 9 + 4 + 1

11. 7 + 5 + 3

12. 2 + 8 + 1

13. 6 + 0 + 4

14. 8 + 5 + 5

15. 1 + 0 + 9

16. 3 + 7 + 5 + 2

17. 4 + 9 + 1 + 2

18. 2 + 8 + 4 + 3

19. 4 + 6 + 2 + 1

20. 3 + 2 + 7 + 6

21. 9 + 2 + 2 + 1

22. 5 + 2 + 4 + 6

23. 1 + 7 + 3 + 1

24. 1 + 8 + 4 + 2

25. 6 + 3 + 1 + 4

Starter activity 9

Objective
Understand that addition can be done in any order, and that more than two numbers can be added.

Strategies
● Revise the pairs of numbers that total 10: 5 + 5, 6 + 4, 7 + 3, 8 + 2, 9 + 1 and reversals.
● Make sure the children appreciate that although it is an efficient strategy to add pairs that make 10 first, any ordering of the numbers will give the same answer.

Answers

1. 13
2. 11
3. 14
4. 13
5. 11
6. 12
7. 19
8. 17
9. 18
10. 14
11. 15
12. 11
13. 10
14. 18
15. 10
16. 17
17. 16
18. 17
19. 13
20. 18
21. 14
22. 17
23. 12
24. 15
25. 14

100 square

Answers

1. 18
2. 12
3. 15
4. 19
5. 25
6. 21
7. 27
8. 23
9. 26
10. 32
11. 34
12. 48
13. 50
14. 61
15. 56
16. 43
17. 63
18. 59
19. 37
20. 46

Starter activity 10

Resources
A 1–100 square (enlarged from photocopiable page 77).

Objective
Choose and use appropriate calculation strategies to solve problems (such as complementary addition, counting up in ones and then in tens).

Strategies
- Ask the children to take 86 from 100.
- Show them two ways to move from 86 to 100 on the 1–100 square: move across to 90 and then down to 100, or move down to 96 and then across to 100 (see figure below).

1. $100 - 82$
2. $100 - 88$
3. $100 - 85$
4. $100 - 81$
5. $100 - 75$
6. $100 - 79$
7. $100 - 73$
8. $100 - 77$
9. $100 - 74$
10. $100 - 68$

11. $100 - 66$
12. $100 - 52$
13. $100 - 50$
14. $100 - 39$
15. $100 - 44$
16. $100 - 57$
17. $100 - 37$
18. $100 - 41$
19. $100 - 63$
20. $100 - 54$

74	75	76	77	78	79	80
84	85	86	87	88	89	90
94	95	96	97	98	99	100

Target totals

1. 10

2. 11

3. 17

4. 21

5. 15

6. 2

7. 4

8. 9

9. 12

10. 18

11. 24

12. 5

Starter activity 11

Resources
A board or flip chart.

Objective
Know by heart all addition and subtraction facts for each number to 20.

Strategies
● Write the numbers 6 8 3 7 well spaced on the board.
● Ask the class to use some or all of these numbers with addition and/or subtraction to make given totals (eg 16 = 7 + 6 + 3).

Answers

More than one answer is often possible.

1. 3 + 7
2. 8 + 3
3. 8 + 3 + 6
4. 6 + 8 + 7
5. 8 + 7
6. 8 − 6
7. 7 − 3
8. 8 − 6 + 7
9. 7 − 3 + 8
10. 7 + 3 + 8
11. 8 + 7 + 3 + 6
12. 8 − 6 + 3

---

20 questions

Remind the children that 20 is made of two tens. So if 5 + 5 = 10 then 15 + 5 must equal 20.

Ask individual children to write pairs of numbers that make 20 on the board. Use this to emphasise commutativity (eg 18 + 2 = 2 + 18).

Point out that the units in each pair will total 10, with the exception of three pairs. Can the children identify them? (20 + 0, 0 + 20, 10 + 10)

Starter activity 12

Resources
A board or flip chart.

Objective
Know by heart all addition facts for each number to 20.

Strategies
● Quickly revise the pairs of numbers that make 10 by saying one of each pair, with the children together saying the other number: 4 (6), 9 (1), 2 (8), 5 (5), 7 (3), 1 (9), 6 (4), 0 (10), 3 (7), 8 (2).

Quick adding

Answers

1. 4
2. 10
3. 4
4. 10
5. 20
6. 14
7. 18
8. 8
9. 7
10. 10
11. 10
12. 17
13. 17
14. 13
15. 16
16. 19
17. 15
18. 13
19. 17
20. 13
21. 19
22. 20
23. 16
24. 15
25. 18
26. 14
27. 17
28. 14
29. 18
30. 19

Starter activity 13

Objective
Know by heart all addition facts for each number to 20.

Strategies
• This is a rapid recall session for pairs of numbers with totals up to 20. Children raise their hands to answer individually.

1. 2 + 2
2. 5 + 5
3. 3 + 1
4. 6 + 4
5. 10 + 10
6. 13 + 1
7. 16 + 2
8. 4 + 4
9. 5 + 2
10. 7 + 3
11. 8 + 2
12. 14 + 3
13. 15 + 2
14. 5 + 8
15. 9 + 7

16. 7 + 12
17. 8 + 7
18. 7 + 6
19. 8 + 9
20. 9 + 4
21. 13 + 6
22. 6 + 14
23. 11 + 5
24. 3 + 12
25. 4 + 14
26. 9 + 5
27. 13 + 4
28. 6 + 8
29. 12 + 6
30. 11 + 8

Quick subtraction

1. 4 – 2

2. 8 – 1

3. 19 – 1

4. 19 – 2

5. 6 – 5

6. 6 – 6

7. 10 – 5

8. 10 – 9

9. 8 – 4

10. 5 – 3

11. 12 – 8

12. 11 – 6

13. 18 – 6

14. 12 – 5

15. 9 – 3

16. 13 – 7

17. 11 – 4

18. 17 – 8

19. 9 – 4

20. 12 – 7

21. 13 – 6

22. 19 – 13

23. 15 – 6

24. 14 – 8

25. 12 – 6

26. 15 – 7

27. 19 – 8

28. 16 – 9

29. 13 – 5

30. 18 – 11

Starter activity 14

Objective
Know by heart all subtraction facts for each number to 20.

Strategies
• This is a rapid recall session for subtraction facts with numbers up to 20.

Answers

1. 2
2. 7
3. 18
4. 17
5. 1
6. 0
7. 5
8. 1
9. 4
10. 2
11. 4
12. 5
13. 12
14. 7
15. 6
16. 6
17. 7
18. 9
19. 5
20. 5
21. 7
22. 6
23. 9
24. 6
25. 6
26. 8
27. 11
28. 7
29. 8
30. 7

Double time

Answers

1. 22
2. 26
3. 24
4. 30
5. 28
6. 22
7. 26
8. 30
9. 24
10. 28
11. 12
12. 15
13. 13
14. 11
15. 14

Starter activity 15

Resources
A board or flip chart.

Objective
Derive quickly doubles of whole numbers to 15 and corresponding halves.

Strategies
• Write 11 12 13 14 15 well spaced on the board. Ask for volunteers to write the doubles of these numbers beneath them.
• Say together: 'Double 11 is 22, double 12...' and so on.
• Introduce the idea of corresponding halves, saying 'Half 22 is 11, half 24...' and so on.
• Ask individual children to complete each statement.

1. double 11 is...
2. twice 13 is...
3. 2 times 12 is...
4. double 15 is...
5. 14 plus 14 is...
6. twice 11 is...
7. 13 plus 13 is...
8. 2 times 15 is...
9. double 12 is...
10. twice 14 is...
11. half 24 is...
12. half 30 is...
13. half 26 is...
14. half 22 is...
15. half 28 is...

Number families

Answers

Two addition and two subtraction statements for each number family, eg
9 + 6 = 15
6 + 9 = 15
15 − 9 = 6
15 − 6 = 9

Starter activity 16

Resources
A board or flip chart.

Objective
Extend understanding that subtraction is the inverse of addition.

Strategies
• Write 6 7 13 on the board.
• Ask for two addition statements using all of these numbers.
• Ask for two subtraction statements using all of the numbers.
• Ask for all four statements using each of the number families shown.

1. 9 15 6
2. 18 12 6
3. 17 8 25
4. 10 50 40
5. 9 26 35
6. 13 21 8
7. 16 27 11
8. 25 13 12
9. 20 30 50
10. 19 5 24
11. 9 19 28
12. 33 4 29
13. 7 21 14
14. 15 23 8
15. 12 19 31

SCHOLASTIC

Money problems

1. I have 3 coins with a total value of 17p. What coins do I have?

2. I have 3 coins with a total value of 23p...

3. I have 3 coins with a total value of 31p...

4. I have 3 coins with a total value of 8p...

5. I have 3 coins with a total value of 53p...

6. I have 3 coins with a total value of 80p...

7. I have 3 coins with a total value of 35p...

8. I have 3 coins with a total value of 57p...

9. I have 3 coins with a total value of 71p...

10. I have 3 coins with a total value of 65p...

11. I have 3 coins with a total value of 27p...

12. I have 3 coins with a total value of 75p...

13. I have 4 coins with a total value of 18p...

14. I have 4 coins with a total value of 85p...

15. I have 4 coins with a total value of 36p...

16. I have 4 coins with a total value of 58p...

17. I have 4 coins with a total value of 82p...

18. I have 4 coins with a total value of 70p...

19. I have 4 coins with a total value of 66p...

20. I have 5 coins with a total value of 37p...

Starter activity 17

Resources
Large pictures of coins to 50p displayed for the class to see (or one of each coin between two, with children working in pairs).

Objective
Solve simple word problems involving money (deciding which coins to use).

Strategies
● Encourage the strategy of working out the highest value coin first.

Answers

1. 10p, 5p, 2p
2. 20p, 2p, 1p
3. 20p, 10p, 1p
4. 5p, 2p, 1p
5. 50p, 2p, 1p
6. 50p, 20p, 10p
7. 20p, 10p, 5p
8. 50p, 5p, 2p
9. 50p, 20p, 1p
10. 50p, 10p, 5p
11. 20p, 5p, 2p
12. 50p, 20p, 5p
13. 10p, 5p, 2p, 1p
14. 50p, 20p, 10p, 5p
15. 20p, 10p, 5p, 1p
16. 50p, 5p, 2p, 1p
17. 50p, 20p, 10p, 2p
18. 50p, 10p, 5p, 5p **or** 20p, 20p, 20p, 10p
19. 50p, 10p, 5p, 1p
20. 10p, 10p, 10p, 5p, 2p **or** 20p, 10p, 5p, 1p, 1p

Money problems

Answers

1. 20p
2. 6p
3. 10p
4. 24p
5. 16p
6. 22p
7. 8p
8. 26p
9. 14p
10. 28p
11. 18p
12. 30p

13. 10p
14. 50p
15. 30p
16. 40p
17. 20p
18. 45p
19. 15p
20. 25p
21. 6p

22. 15p
23. 30p
24. 9p
25. 12p

Starter activity 18

Resources
Large pictures of coins to 50p displayed for the class to see.

Objective
Solve simple word problems involving money (finding the cost of more than one item).

Strategies
• Remind the children that to find the cost of two items, we double (×2) the cost of one item.

• Ask the children how they would find the cost of five items. Establish that they should multiply by 5.

• Ask the children to find the cost of three items. Explain that this could be done by adding or by multiplying.

1. I costs 10p... 2 will cost?
2. I costs 3p...
3. I costs 5p...
4. I costs 12p...
5. I costs 8p...
6. I costs 11p...

7. I costs 4p...
8. I costs 13p...
9. I costs 7p...
10. I costs 14p...
11. I costs 9p...
12. I costs 15p...

13. I costs 2p... 5 will cost?
14. I costs 10p...
15. I costs 6p...
16. I costs 8p...

17. I costs 4p...
18. I costs 9p...
19. I costs 3p...
20. I costs 5p...

21. I costs 2p... 3 will cost?
22. I costs 5p...
23. I costs 10p...

24. I costs 3p...
25. I costs 4p...

Money problems

1. How much do two pencils cost if each pencil costs 12p?

2. I have 7p and you have 8p. How much do we have altogether?

3. I had 20p, then spent 17p. How much do I have left?

4. Two lollies cost 14p altogether. How much was each lolly?

5. I have six identical coins. I have 12p. What coins do I have?

6. I had 40p, but then lost 10p. How much do I have left?

7. Matthew had two 20p coins and one 5p coin. How much did he have?

8. Michael had three 10p coins and one 20p coin. How much did he have?

9. If I saved 10p each week, how long would it take me to save 50p?

10. If I had one of every coin from 1p to 50p, how much would I have?

11. I have 12p and you have 15p. How much do we have altogether?

12. I had 25p, then spent 18p. How much do I have left?

13. Pencils cost 14p each. I wanted two, but only had 20p. How much more did I need?

14. My friend and I had 50p between us. I had 28p. How much did he have?

15. I saved 25p each week for 3 weeks. How much had I saved?

Starter activity 19

Objective
Solve simple word problems involving money.

Strategies
● Allow time for less able children to work out the answer.
● Repeating the question, using different wording, may be helpful.

Answers

1. 24p
2. 15p
3. 3p
4. 7p
5. 2p coins
6. 30p
7. 45p
8. 50p
9. 5 weeks
10. 88p
11. 27p
12. 7p
13. 8p
14. 22p
15. 75p

Addition grid

Answers

1. even
2. odd
3. even
4. odd
5. odd
6. odd
7. even
8. even
9. odd
10. even
11. odd
12. even

Starter activity 20

Resources
One large prepared addition grid.

Objective
Use patterns of similar calculations to explore addition with odd and even numbers.

Strategies
• Ask targeted individuals to fill in the cells.
• Make sure the children understand that all the answers in this grid are odd numbers.

+	1	3	5	7	9
2					
4					
6					
8					
10					

• Encourage general statements about adding odd and even numbers.

odd + even = odd

even + even = even

odd + odd = even

• Ask children to say whether the answers to these addition questions are odd or even.

1. 3 + 5

2. 14 + 15

3. 10 + 12

4. 16 + 19

5. 18 + 11

6. 17 + 12

7. 9 + 11

8. 6 + 12

9. 13 + 14

10. 15 + 13

11. 10 + 11

12. 19 + 13

Addition grid

+	2		7
6		9	
7			
			19

When the grid is complete, ask the children to work in pairs, writing as many addition and subtraction statements from the grid as they can. Write examples on the board (see below).

$$2 + 6 = 8 \qquad 6 + 2 = 8$$

$$8 - 2 = 6 \qquad 8 - 6 = 2$$

Target 50

$$30 + 20 = 50 \qquad 17 + 33 = 50$$

$$29 + 21 = 50 \qquad 13 + 37 = 50$$

Answers

1. 5p, 10p,
    20p, 50p
2. 2
3. 4
4. 6
5. 10
6. 2
7. 4
8. 6
9. 10
10. 5
11. 10
12. 20p, 20p,
    10p
13. 2
14. 4
15. 10
16. 20
17. 20p, 20p,
    10p, 10p
18. 20p, 10p,
    10p, 5p, 5p
19. 50p, 20p,
    5p
20. £1, 20p,
    20p, 5p

Starter activity 22

Resources
Large pictures of silver coins and the £1 coin.

Objective
Recognise all coins.

Strategies
● Tell the children that they can use only silver coins and the £1 coin to answer these questions.

Silver start

1. Name the four silver coins.

2. How many 50p coins are equal to a £1 coin?

3. How many 50p coins equal £2?

4. How many 50p coins equal £3?

5. How many 50p coins equal £5?

6. How many 10p coins are equal to one 20p?

7. How many 10p coins equal two 20p coins?

8. How many 10p coins equal three 20p coins?

9. How many 10p coins are worth the same as a £1 coin?

10. How many 20p coins are worth the same as a £1 coin?

11. How many 20p coins equal £2?

12. Which three silver coins equal 50p?

13. How many 5p coins equal 10p?

14. How many 5p coins equal 20p?

15. How many 5p coins equal 50p?

16. How many 5p coins are there in £1?

17. Make 50p using four coins.

18. Make 50p using five silver coins.

19. Which three coins total 75p?

20. Which four coins total £1.45?

Cash count

1. Which coin is worth the most?

2. How many different bronze coins are there?

3. What are they?

4. If you had the three most valuable coins, how much would you have?

5. What are the silver coins?

6. What are all the different silver coins worth together?

7. If you had one of each coin, how much would you have?

8. How many 50p coins equal £1?

9. How many 50p coins equal £2?

10. How many 20p coins equal £1?

11. How many 20p coins equal £2?

12. How many 1p coins equal £1?

13. How many 10p coins equal £1?

14. How many 5p coins equal £1?

15. How many 2p coins equal 10p?

Objective
Recognise all coins.

Strategies
● Count together in 20s to 500, then in 50s to 500.
● Ask the children to name all the coins in order of increasing value.

1. the £2 coin
2. 2
3. 2p and 1p
4. £3.50
5. 50p, 20p, 10p, 5p
6. 85p
7. £3.88
8. 2
9. 4
10. 5
11. 10
12. 100
13. 10
14. 20
15. 5

Multiples of 2

Starter activity 25

Resources
A board or flip chart.

Objective
Know by heart multiplication facts for the 2 times table.

Strategies
• Draw a 'clock face' on the board with the numbers 1–10 arranged randomly in a circle. In the centre, write '×2'.
• The children face the board and, in unison, say each multiplication fact three times in succession. This will encourage children who are unsure the first time to join in the repetitions.

8

4   3

7

6

×2

10

2

1

9

5

Multiples of 4

Answers

1. 8
2. 24
3. 4
4. 32
5. 12
6. 36
7. 16
8. 40
9. 28
10. 20

Starter activity 26

Resources
A board or flip chart.

Objective
Begin to know the 4 times table.

Strategies
• Build up the 4 times table on the board. Look at the units pattern. Notice that all answers are even numbers.
• Say the table together several times. Count in 4s forwards and backwards.
• Ask individuals to answer questions while the table can be seen.
• Erase the better-known answers, then try again.

1. 2 × 4

2. 6 × 4

3. 1 × 4

4. 8 × 4

5. 3 × 4

6. 9 × 4

7. 4 × 4

8. 10 × 4

9. 7 × 4

10. 5 × 4

Magic square

	5	

One solution is shown below.
Many others are possible.

6	7	2
1	5	9
8	3	4

Starter activity 27

Resources
A board or flip chart; paper and pencils for each pair.

Objective
Solve mathematical problems or puzzles, recognise simple patterns and relationships, generalise and predict.

Strategies
• Divide the class into mixed-ability pairs.
• Draw the grid shown above left on the board. Ask the children to copy it and then to create a 'magic' square in which each row, column and diagonal gives a total of 15. They should use the numbers 1–9 once only.

Steps of 5

```
0   1   2   3   4   5   6   7   8   9   10
|   |   |   |   |   |   |   |   |   |   |
0   5   10
```

1. 2 times 5 equals
2. 8 multiplied by 5 equals
3. three 5s are
4. 7 steps of 5 make
5. How many 5s make 20?
6. 10 times 5 equals
7. 5 times itself is
8. 1 multiplied by 5 equals
9. the product of 6 and 5 is
10. 9 times 5 is
11. double 5
12. 8 times 5
13. 6 times 5
14. 9 multiplied by 5
15. the product of 7 and 5 is

Starter activity 28

Resources
A number line drawn on the board or flip chart.

Objective
Recognise two-digit multiples of 5.

Strategies
• Ask individual children to write multiples of 5 under the line.
• When the line is complete, count in 5s forwards and backwards. Say the 5 times table together. Point to numbers out of sequence.
• Remove the number line and ask quick-fire questions.

Answers

1. 10
2. 40
3. 15
4. 35
5. 4
6. 50
7. 25
8. 5
9. 30
10. 45
11. 10
12. 40
13. 30
14. 45
15. 35

Multiples of 2, 5 and 10

Answers

1. 8
2. 25
3. 30
4. 100
5. 30
6. 16
7. 4
8. 15
9. 70
10. 50
11. 20
12. 20
13. 50
14. 18
15. 5
16. 40
17. 6
18. 40
19. 10
20. 12
21. 35
22. 20
23. 80
24. 45
25. 2
26. 10
27. 14
28. 60
29. 10
30. 90

Starter activity 29

Objective
Recognise two-digit multiples of 2, 5 and 10.

Strategies
• Count in 2s, 5s and 10s with the children.
• Remind the children of the patterns in each table, eg all multiples of 2 are even numbers; 5× table products end alternately in 5 and 0.

1. 4×2
2. 5×5
3. 3×10
4. 10×10
5. 6×5
6. 8×2
7. 2×2
8. 3×5
9. 7×10
10. 10×5
11. 4×5
12. 2×10
13. 5×10
14. 9×2
15. 1×5

16. 4×10
17. 3×2
18. 8×5
19. 1×10
20. 6×2
21. 7×5
22. 10×2
23. 8×10
24. 9×5
25. 1×2
26. 2×5
27. 7×2
28. 6×10
29. 5×2
30. 9×10

SCHOLASTIC

Derive doubles

1. double 20

2. add 16 to itself

3. double 18

4. double 17

5. 2 times 19

6. twice 18

7. double 16

8. twice 20

9. add 19 to itself

10. 2 times 17

11. double 10

12. twice 8

13. add 6 to itself

14. 2 times 15

15. double 9

16. 2 times 11

17. twice 13

18. double 12

19. 2 times 14

20. twice 7

21. double 16

22. double 19

23. double 17

24. double 20

25. double 18

Starter activity 30

Answers

Resources
A board or flip chart.

Objective
Derive quickly doubles of all whole numbers to at least 20.

Strategies
● Ask the children for the doubles of 11, 12, 13, 14, 15 learned previously.
● Write the numbers 16 17 18 19 20 on the board, well spaced. Ask individuals to write the double of each number beneath it and to explain how they worked it out.
● Practise the examples as a class, using different 'doubles' language.

● Recap on known doubles.

1. 40
2. 32
3. 36
4. 34
5. 38
6. 36
7. 32
8. 40
9. 38
10. 34

11. 20
12. 16
13. 12
14. 30
15. 18
16. 22
17. 26
18. 24
19. 28
20. 14
21. 32
22. 38
23. 34
24. 40
25. 36

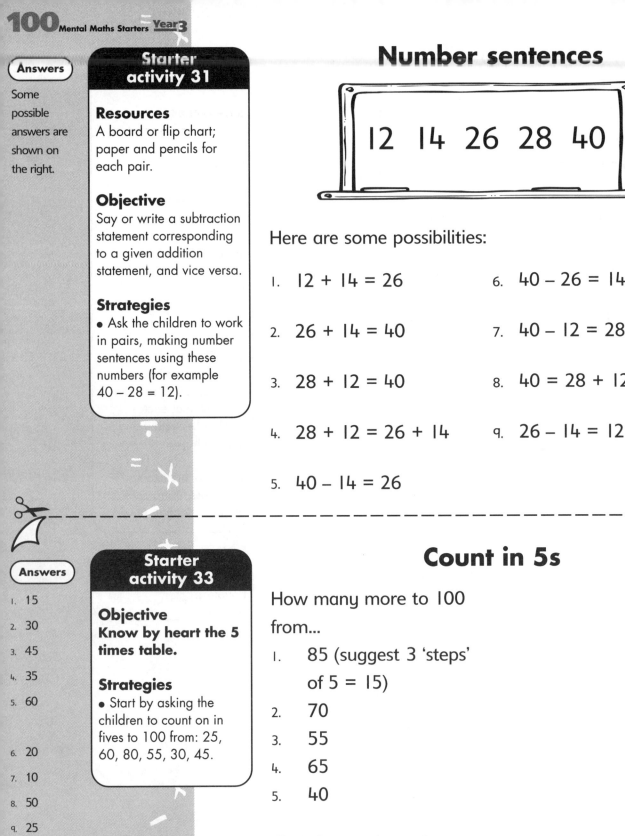

Number sentences

12 14 26 28 40

Here are some possibilities:

1. 12 + 14 = 26

2. 26 + 14 = 40

3. 28 + 12 = 40

4. 28 + 12 = 26 + 14

5. 40 – 14 = 26

6. 40 – 26 = 14

7. 40 – 12 = 28

8. 40 = 28 + 12

9. 26 – 14 = 12

Answers

Some possible answers are shown on the right.

Starter activity 31

Resources
A board or flip chart; paper and pencils for each pair.

Objective
Say or write a subtraction statement corresponding to a given addition statement, and vice versa.

Strategies
• Ask the children to work in pairs, making number sentences using these numbers (for example 40 – 28 = 12).

Answers

1. 15
2. 30
3. 45
4. 35
5. 60
6. 20
7. 10
8. 50
9. 25
10. 40
11. 30
12. 15
13. 35
14. 45
15. 20

Starter activity 33

Objective
Know by heart the 5 times table.

Strategies
• Start by asking the children to count on in fives to 100 from: 25, 60, 80, 55, 30, 45.

Count in 5s

How many more to 100 from...

1. 85 (suggest 3 'steps' of 5 = 15)

2. 70

3. 55

4. 65

5. 40

Find the product of...

6. 4 × 5

7. 2 × 5

8. 10 × 5

9. 5 × 5

10. 8 × 5

11. 6 and 5

12. 3 and 5

13. 7 and 5

14. 9 and 5

15. 4 and 5

More doubles

What is double...?

1. 10

2. 7

3. 12

4. 5

5. 13

What is half of...?

6. 20

7. 14

8. 24

9. 10

10. 6

What is double...?

11. 6

12. 2

13. 14

14. 18

15. 11

What is half of...?

16. 8

17. 12

18. 30

19. 36

20. 22

Starter activity 32

Objective
Use doubling or halving to multiply or divide, starting from known facts.

Strategies
● Remind the children that doubles have corresponding halves, for example double 8 (8 × 2) is 16, half of 16 (16 ÷ 2) is 8.
● Practise known doubles and halves with a quickfire session.

Answers

1. 20
2. 14
3. 24
4. 10
5. 26

6. 10
7. 7
8. 12
9. 5
10. 3

11. 12
12. 4
13. 28
14. 36
15. 22

16. 4
17. 6
18. 15
19. 18
20. 11

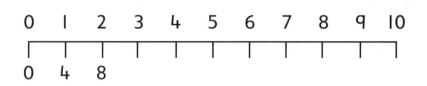

Count in 4s

Starter activity 34

Resources
A number line drawn on the board, a pointer.

Objective
Begin to know the 4 times table.

Strategies
• Ask individual children to write the multiples of 4 under the line.
• Practise counting in 4s. Say the 4 times table in order. Point to numbers out of sequence.
• Ask the whole class questions.

```
0   1   2   3   4   5   6   7   8   9   10
|   |   |   |   |   |   |   |   |   |   |
0   4   8
```

1. 3 times 4

2. 5 multiplied by 4

3. 7 times 4

4. twice 4 is

5. I group of 4

6. 9 multiplied by 4

7. 4 times 4

8. the product of 6 and 4 is

9. 10 times 4

10. 8 multiplied by 4

Starter activity 35

Resources
A pencil and a 1–100 square (photocopiable page 77) for each child.

Objective
Begin to know the 3 times table.

Strategies
• The activity is described in the text on the right.
• Individuals may volunteer to try counting in 3s alone, with others raising a hand if a mistake is made.

Steps of 3

Ask the children to draw a line through all the multiples of 3 up to 60 on the 1–100 square.

When they have finished, count together in 3s to 60 and back again. How far can the children go without looking at their 1–100 squares?

Write the 3 times table to 30 on the board. Ask the class to say it together.

■SCHOLASTIC

Table talk

1. 2×5
2. 7×2
3. 8×10
4. 6×2
5. 9×5
6. 5×5
7. 9×2
8. 6×10
9. 7×5
10. 1×5

11. 4×3
12. 8×2
13. 5×10
14. 6×5
15. 7×10
16. 5×4
17. 4×2
18. 10×3
19. 3×4
20. 8×5

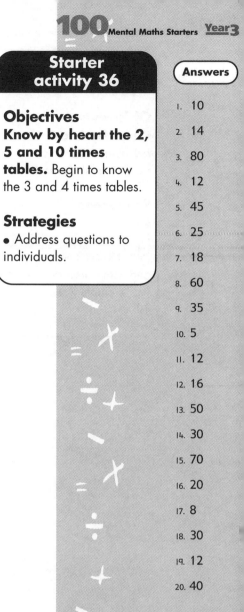

Starter activity 36

Answers

Objectives
Know by heart the 2, 5 and 10 times tables. Begin to know the 3 and 4 times tables.

Strategies
● Address questions to individuals.

1. 10
2. 14
3. 80
4. 12
5. 45
6. 25
7. 18
8. 60
9. 35
10. 5
11. 12
12. 16
13. 50
14. 30
15. 70
16. 20
17. 8
18. 30
19. 12
20. 40

Addition bonds

Answers

1. 6
2. 14
3. 20
4. 6
5. 9
6. 15
7. 12
8. 14
9. 16
10. 9
11. 16
12. 19
13. 18
14. 12
15. 14
16. 12
17. 9
18. 13
19. 17
20. 15
21. 16
22. 16
23. 13
24. 12
25. 19
26. 18
27. 17
28. 13
29. 14
30. 20

Starter activity 37

Objective
Know by heart all addition facts for each number to 20.

Strategies
• Emphasise counting on from the larger number (5 + 13 = 13 + 5 = 18) and using near doubles (for example 7 + 6 = 12 + 1 = 13).

1. 3 + 3
2. 10 + 4
3. 8 + 12
4. 6 + 0
5. 5 + 4
6. 7 + 8
7. 9 + 3
8. 3 + 11
9. 1 + 15
10. 6 + 3
11. 4 + 12
12. 17 + 2
13. 9 + 9
14. 6 + 6
15. 7 + 7

16. 4 + 8
17. 0 + 9
18. 6 + 7
19. 9 + 8
20. 2 + 13
21. 7 + 9
22. 10 + 6
23. 8 + 5
24. 3 + 9
25. 14 + 5
26. 1 + 17
27. 12 + 5
28. 2 + 11
29. 8 + 6
30. 11 + 9

■SCHOLASTIC

Subtraction bonds

1. $5 - 2$

2. $8 - 7$

3. $4 - 1$

4. $6 - 0$

5. $10 - 7$

6. $9 - 2$

7. $12 - 2$

8. $7 - 3$

9. $6 - 4$

10. $4 - 4$

11. $7 - 6$

12. $9 - 5$

13. $12 - 10$

14. $11 - 7$

15. $3 - 3$

16. $8 - 0$

17. $14 - 9$

18. $17 - 7$

19. $10 - 4$

20. $16 - 13$

21. $18 - 9$

22. $14 - 7$

23. $6 - 3$

24. $17 - 11$

25. $15 - 8$

26. $7 - 4$

27. $10 - 3$

28. $8 - 5$

29. $9 - 9$

30. $13 - 8$

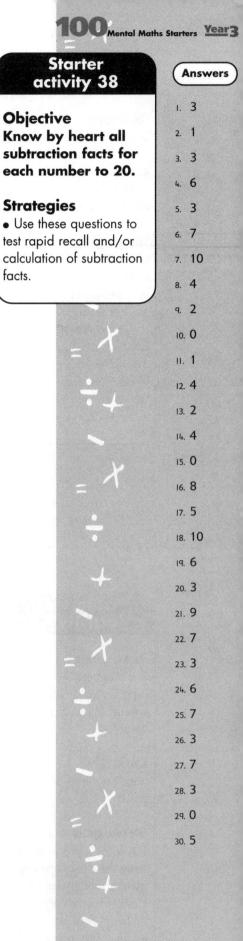

Starter activity 38

Answers

Objective
Know by heart all subtraction facts for each number to 20.

Strategies
● Use these questions to test rapid recall and/or calculation of subtraction facts.

1. 3
2. 1
3. 3
4. 6
5. 3
6. 7
7. 10
8. 4
9. 2
10. 0
11. 1
12. 4
13. 2
14. 4
15. 0
16. 8
17. 5
18. 10
19. 6
20. 3
21. 9
22. 7
23. 3
24. 6
25. 7
26. 3
27. 7
28. 3
29. 0
30. 5

Addition bonds

1. 25
2. 24
3. 20
4. 24
5. 22
6. 33
7. 36
8. 31
9. 21
10. 21
11. 23
12. 22
13. 25
14. 34
15. 33
16. 32
17. 34
18. 42
19. 41
20. 42

Starter activity 39

Resources
A board or flip chart.

Objective
Bridge through a multiple of 10, then adjust.

Strategies
● Write 16 + 7 = on the board. Ask for the answer and strategies to find it.
● If necessary, suggest 'bridging the 10' or using doubles:
16 + 7 = 16 + 4 + 3
= 20 + 3 = 23
or 16 + 7 = 16 + 6 + 1
= 10 + 12 + 1 = 23

1. 18 + 7
2. 16 + 8
3. 14 + 6
4. 19 + 5
5. 15 + 7
6. 27 + 6
7. 29 + 7
8. 26 + 5
9. 17 + 4
10. 15 + 6

11. 18 + 5
12. 16 + 6
13. 19 + 6
14. 28 + 6
15. 26 + 7
16. 29 + 3
17. 27 + 7
18. 38 + 4
19. 36 + 5
20. 39 + 3

Doubles revision

1. 15
2. 5
3. 20
4. 16
5. 10
6. 2
7. 9
8. 24
9. 12
10. 20

Starter activity 41

Resources
The numbers 0–10 written on a board or OHT, well spread out and at child height.

Objective
Recall doubles of numbers to 10.

Strategies
● Ask for a volunteer to write the double of any number underneath it. Continue until they have all been doubled.
● The children work in pairs for three minutes, practising these doubles.
● Ask questions, with children raising their hands.

1. double 7 add 1
2. double 3 minus 1
3. double 8 plus 4
4. double 4 plus double 4
5. take 4 from double 7

6. double 1 minus 0
7. take 3 from double 6
8. double 9, add 6
9. double 2 plus 8
10. double 5 plus double 5

■ SCHOLASTIC

Addition and subtraction facts

Make an addition fact from:

1. $24 - 19 = 5$

2. $69 - 65 = 4$

3. $40 - 32 = 8$

4. $73 - 69 = 4$

5. $85 - 78 = 7$

6. $102 - 97 = 5$

7. $106 - 101 = 5$

8. $131 - 127 = 4$

9. $148 - 142 = 6$

10. $107 - 103 = 4$

Make a subtraction fact from:

11. $15 + 8 = 23$

12. $22 + 14 = 36$

13. $17 + 18 = 35$

14. $36 + 24 = 60$

15. $23 + 49 = 72$

Starter activity 40

Resources
A board or flip chart.

Objective
Understand that subtraction is the inverse of addition.

Strategies
• Write $19 + 7 =$ and $32 - 28 =$ on the board.
• Ask for volunteers to write an addition fact and a subtraction fact from each of the two examples on the board.
• Stress the usefulness of 'counting up' when subtracting numbers that are close together.
• It may be necessary to write the larger numbers on the board when working through these examples.

Answers

1. $19 + 5 = 24$ (**or** $5 + 19 = 24$)

2. $65 + 4 = 69$

3. $32 + 8 = 40$

4. $69 + 4 = 73$

5. $78 + 7 = 85$

6. $97 + 5 = 102$

7. $101 + 5 = 106$

8. $127 + 4 = 131$

9. $142 + 6 = 148$

10. $103 + 4 = 107$

11. $23 - 15 = 8$ (**or** $23 - 8 = 15$)

12. $36 - 22 = 14$

13. $35 - 18 = 17$

14. $60 - 36 = 24$

15. $72 - 49 = 23$

Take away

Starter activity 42

Objective
Understand the operation of subtraction and the related vocabulary.

Strategies
● Use these questions to practise subtraction with the class.

1. Take 9 from 12.

2. Find the difference between 18 and 13.

3. 11 minus 5

4. Find the difference between 8 and 6.

5. How many more is 16 than 7?

6. Take 5 from 14.

7. Subtract 3 from 15.

8. 19 minus 12

9. How much more is 20 than 7?

10. Subtract 8 from 11.

11. Find the difference between 17 and 4.

12. 8 minus 8

13. How much less than 13 is 3?

14. Find the difference between 10 and 1.

15. 5 minus zero

16. Subtract 7 from 18.

17. 19 minus 13

18. How much more is 14 than 6?

19. Take 6 from 17.

20. How much less than 15 is 9?

Work it out

1. I think of a number, then subtract 4. The answer is 6. What was my number?

2. I think of a number, then add 7. The answer is 16. What was my number?

3. My friend and I together spend 50p. He spent 30p. How much did I spend?

4. A spider has 8 legs. How many legs do 2 spiders have?

5. How many legs would 5 spiders have?

6. Insects have 6 legs. How many legs would 3 ladybirds have?

7. How many legs would 10 ladybirds have?

8. I have £2 to give to four children. They will each have the same amount. How much will that be?

9. Peter saved 50p a week for 6 weeks. How much did he save?

10. 80 crayons are put into packets that hold 10 each. How many packets will be needed?

Starter activity 43

Objective
Choose and use appropriate operations and ways of calculating **to solve problems.**

Strategies
• Discuss some of these problems, asking for the strategies used.

Answers
1. 10
2. 9
3. 20p
4. 16
5. 40
6. 18
7. 60
8. 50p
9. £3.00
10. 8

Odd man out

The children stand in a line or circle. The first child says the number given. The next child halves that number, and so on. When an odd number is reached, the child who says it is 'out' and sits down. The game begins again, with the next child saying the new starting number.

Starting numbers:

16	20
10	28
8	32
24	12
6	100

Starter activity 44

Objective
Use doubling or halving to multiply or divide.

Strategies
• Use this game to practise halving with the class.

Starter activity 45

Objective
Use doubling or halving to multiply or divide.

Double up

The children stand in a line or circle. The first child says the number given. The next child doubles that number, and so on. When a number greater than 50 is reached, the child who says it is 'out' and sits down. The game begins again, with the next child saying the new number.

Starting numbers:

3	4
11	6
2	7
5	10
9	13

Starter activity 47

Resources
A board or flip chart, one copy of photocopiable page 78 per child, coloured pencils.

Objective
Begin to recognise simple equivalent fractions.

Strategies
● Let the children complete photocopiable page 78.

Number fractions

Draw this shape on the board:

Ask a child to shade in one quarter of the shape.

Now shade another quarter. Ask: *What fraction of the shape is shaded?*

Establish that:

two quarters equal one half

$$\frac{1}{4} + \frac{1}{4} = \frac{1}{2}$$

Half and quarter

What is half of...?

1. 10

2. 16

3. 4

4. 20

5. 60

6. 100

7. 18

8. 6

9. 14

10. 24

What is one quarter of...?

11. 8

12. 20

13. 40

14. 4

15. 12

Starter activity 46

Resources
A board or flip chart.

Objective
Recognise unit fractions such as $\frac{1}{2}$, $\frac{1}{4}$.

Strategies
• Remind the children that to find a half, we divide by 2.

• Write: $\frac{1}{4}$ of 12
$= 12 \div 4 = 3$
or $12 \rightarrow 6 \rightarrow 3$
• Stress that to find a quarter we divide by 4, or we halve twice.

Answers

1. 5
2. 8
3. 2
4. 10
5. 30
6. 50
7. 9
8. 3
9. 7
10. 12

11. 2
12. 5
13. 10
14. 1
15. 3

Pieces of 8

1. 8

2. $\frac{2}{8}$ or $\frac{1}{4}$

3. $\frac{4}{8}$ or $\frac{2}{4}$
   or $\frac{1}{2}$

4. 1

5. $\frac{7}{8}$

6. $\frac{7}{8}$

7. $\frac{3}{8}$

8. $\frac{4}{8}$ or $\frac{2}{4}$
   or $\frac{1}{2}$

Starter activity 48

Resources
A board or flip chart.

Objective
Begin to recognise simple equivalent fractions.

Strategies
• Draw a rectangle divided into eighths. Label one section $\frac{1}{8}$ and invite individuals to label the other seven sections.
• Emphasise that each section is 'one of the eight equal parts'.
• Ask individuals to: fill 2 eighths with 'spots'; fill 4 eighths with 'stripes'; fill 1 eighth with 'curls'.

1. How many eighths are there?
2. What fraction of the shape has 'spots'? (encourage equivalents)
3. What fraction has 'stripes'? (encourage equivalents)
4. How many eighths are blank?
5. If I took away the 'curly' eighth, what fraction of the rectangle would be left?
6. If I took away the blank eighth, what fraction of the rectangle would be left?
7. If I put spots into the blank eighth, what fraction would then be 'spotty'?
8. What fraction would be left if I took away the 'striped' part?

Starter activity 49

Resources
A board or flip chart, one copy of photocopiable page 78 per child, coloured pencils.

Objective
Begin to recognise simple equivalent fractions.

Strategies
• Discuss the task briefly, emphasising the idea of equivalent fractions.
• Allow a short time for the children to complete the sheet, then ask them for some equivalent fractions. List these on the board (see examples on right). Include ¾ and its equivalents if offered.
• Draw three aligned rectangles and divide them into halves, quarters and eighths respectively to highlight the equivalence.

Shape fractions

Examples of equivalent fractions:

$$\frac{1}{4} + \frac{1}{4} = \frac{2}{4} = \frac{1}{2}$$

$$\frac{1}{8} + \frac{1}{8} + \frac{1}{8} + \frac{1}{8} = \frac{4}{8} = \frac{2}{4} = \frac{1}{2}$$

$$\frac{1}{8} + \frac{1}{8} = \frac{4}{8} = \frac{2}{8} = \frac{1}{4}$$

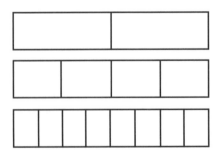

■SCHOLASTIC

2× and 5×

1. 4 × 2

2. 8 × 2

3. How many 2s to 12?

4. 3 × 5

5. 2 × 2

6. 7 × 5

7. How many 5s make 20?

8. 2 × 5

9. 3 × 2

10. How many 2s make 14?

11. 8 × 5

12. 10 × 2

13. 9 × 5

14. How many 5s make 30?

15. 10 × 5

16. 1 × 2

17. How many 2s to 18?

18. 5 × 2

19. 1 × 5

20. How many 5s make 25?

Starter activity 50

Objective
Recognise two-digit multiples of 2 and 5.

Strategies
● Count in 2s and 5s from 0 to 20 and 50 respectively.
● Emphasise the patterns.

Answers
1. 8
2. 16
3. 6
4. 15
5. 4
6. 35
7. 4
8. 10
9. 6
10. 7
11. 40
12. 20
13. 45
14. 6
15. 50
16. 2
17. 9
18. 10
19. 5
20. 5

More 2x

Answers

1. 6
2. 14
3. 8
4. 2
5. 12
6. 20
7. 4
8. 18
9. 10
10. 16
11. 3
12. 7
13. 10
14. 1
15. 9
16. 5
17. 2
18. 8
19. 6
20. 4

Starter activity 51

Objective
Know by heart multiplication facts for the 2 times table.
Derive corresponding division facts.

Strategies
- A rapid recall session.

1. 3×2
2. 7×2
3. 4×2
4. 1×2
5. 6×2
6. 10×2
7. 2×2
8. 9×2
9. 5×2
10. 8×2

11. How many 2s make 6?
12. How many 2s make 14?
13. How many 2s make 20?
14. How many 2s make 2?
15. How many 2s make 18?
16. How many 2s make 10?
17. How many 2s make 4?
18. How many 2s make 16?
19. How many 2s make 12?
20. How many 2s make 8?

Starter activity 52

Resources
A board or flip chart, a pointer.

Objective
Know by heart multiplication facts for the 5 times table.

More 5x

Write the 5 times table on the board. Ask the children to recite it together. Repeat it twice, then remove it. Draw a 'clock face' as in Starter activity 25. The children together say each fact twice. Use the pointer to select at random, increasing in speed. Ask children to give the answer only.

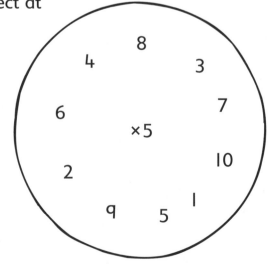

Tens and hundreds

1. 5×10

2. 10×10

3. 6×10

4. 2×10

5. 9×10

6. 4×10

7. 7×10

8. 1×10

9. 8×10

10. 3×10

11. I say 600. You say...?

12. I say 400. You say...?

13. I say 800. You say...?

14. I say 500. You say...?

15. I say 200. You say...?

16. I say 400. You say...?

17. I say 900. You say...?

18. I say 300. You say...?

19. I say 100. You say...?

20. I say 700. You say...?

Starter activity 53

Objectives
Know by heart:
multiplication facts
for the 10 times
table; all pairs of
multiples of 100 with a
total of 1000
(eg 300 + 700).

Strategies
● Count forwards and
backwards in 10s to 100
and in 100s to 1000.

● Remind the children
that 3 + 7 = 10, so 300
+ 700 = 1000.
● Play 'I say, You say'.
The children together
give the other half of a
pair to total 1000.

Answers

1. 50
2. 100
3. 60
4. 20
5. 90
6. 40
7. 70
8. 10
9. 80
10. 30
11. 400
12. 600
13. 200
14. 500
15. 800
16. 600
17. 100
18. 700
19. 900
20. 300

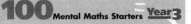

2 and 5 times

Starter activity 54

Objectives
Know by heart multiplication facts for the 2 and 5 times tables. Derive corresponding division facts.

Strategies
• A rapid recall session.

1. 3 × 5
2. 6 multiplied by 2
3. 1 × 2
4. How many 5s make 30?
5. 20 ÷ 5
6. 10 × 5
7. 8 ÷ 2
8. How many 2s make 20?
9. 9 × 5
10. 5 × 2
11. 1 × 5
12. 14 ÷ 2
13. 25 ÷ 5
14. double 2
15. 9 × 2
16. How many 5s make 40?
17. double 5
18. How many 2s make 6?
19. 7 × 5
20. 8 × 2

Combinations

Starter activity 55

Resources
A board or flip chart.

Objective
Solve mathematical problems or puzzles, recognise simple patterns and relationships, generalise and predict. Support extensions by asking 'What if.....?'

Strategies
•Write the numbers 2, 3 and 6 on the board.

Use the three numbers to:

1. give a multiplication fact
2. give another multiplication fact
3. give a division fact
4. give another division fact
5. make 11
6. make 12
7. make 15
8. make 7
9. Is it possible to make 20?
10. Is it possible to make 36?
11. Is it possible to make 1?
12. Is it possible to make 5?
13. Is it possible to make 18?
14. Is it possible to make 24?
15. Is it possible to make 30?

Addition

Write: 18 + 25 =   27 + 27 =

Allow the children a minute to work out the answers.

Discuss the methods used. Gather (or if necessary, suggest) methods such as: beginning with the larger number, then using complementary addition; partitioning; using doubles; dealing with units first.

If time allows, write the following and discuss methods:

1. 16 + 36 =   3. 35 + 26 =
2. 24 + 26 =   4. 23 + 33 =

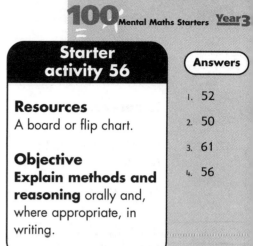

Starter activity 56

Resources
A board or flip chart.

Objective
Explain methods and reasoning orally and, where appropriate, in writing.

Answers

1. 52
2. 50
3. 61
4. 56

Subtraction

Write: 34 – 18 =   42 – 28 =

Allow the children a minute to work out the answers.

Discuss the methods used. Gather (or if necessary, suggest) methods such as: using the nearest 'tens' number, then adjusting; using a near double.

If time allows, write the following and discuss methods:

1. 39 – 19 =   3. 34 – 18 =
2. 46 – 23 =   4. 42 – 29 =

Starter activity 57

Resources
A board or flip chart.

Objective
Explain methods and reasoning orally and, where appropriate, in writing.

Answers

1. 20
2. 23
3. 16
4. 13

Near tens

Answers

1. 32
2. 75
3. 67
4. 31
5. 74
6. 99
7. 87
8. 86
9. 88
10. 93
11. 55
12. 89
13. 72
14. 96
15. 61
16. 54
17. 90
18. 87
19. 73
20. 86

Starter activity 58

Resources
A 1–100 square (copy of photocopiable page 76) per child.

Objective
Add mentally a 'near multiple of 10' to a two-digit number...
by adding 10, 20, 30... and adjusting.

Strategies
• Practise adding multiples of 10, with children using their squares to make 'jumps' of 10.

• Together add 21 to 36. Encourage two jumps of 10 plus 1.

• Try 45 + 31 together.

1. 22 + 10
2. 65 + 10
3. 47 + 20
4. 11 + 20
5. 34 + 40
6. 79 + 20
7. 57 + 30
8. 26 + 60
9. 48 + 40
10. 83 + 10

11. 34 + 21
12. 68 + 21
13. 51 + 21
14. 75 + 21
15. 40 + 21

16. 23 + 31
17. 59 + 31
18. 76 + 11
19. 62 + 11
20. 45 + 41

■SCHOLASTIC

Near tens

Practise, using 100 squares:

1. 24 + 19

2. 42 + 19

3. 65 + 19

4. 38 + 19

5. 57 + 19

Starter activity 59

Resources
A 1–100 square (copy of photocopiable page 76) per child.

Objective
Add mentally a 'near multiple of 10' to a two-digit number...
by adding 10, 20, 30... and adjusting.

Strategies
• Ask: *What is 36 + 19?* Encourage 36 + 20 – 1.

Answers

1. 43
2. 61
3. 84
4. 57
5. 76

6. 44
7. 72
8. 95
9. 50
10. 59

11. 52
12. 76
13. 81
14. 68
15. 73

Practise without squares, visualising the pattern.

6. 25 + 19

7. 53 + 19

8. 76 + 19

9. 31 + 19

10. 40 + 19

If appropriate, continue:

11. 23 + 29

12. 47 + 29

13. 52 + 29

14. 29 + 39

15. 34 + 39

A choice of strategies

$$38 + 27 = 38 + 20 + 7 = 58 + 2 + 5 = 65$$

(partitioning)

$$38 + 27 = 38 + 30 - 3 = 68 - 3 = 65$$

(rounding up, then compensating down)

Answers

1. 67
2. 93
3. 75
4. 69
5. 79
6. 93
7. 60
8. 84
9. 82
10. 89

Starter activity 60

Resources
A board or flip chart; a 1–100 square (copy of photocopiable page 77) per child.

Objective
Use patterns of similar calculations.

Strategies
• Write 22 + 27 and 38 + 27.
• Ask for strategies, using 100 squares.
• Discuss different strategies (see right).

Try, then discuss:

1. 43 + 24
2. 57 + 36
3. 26 + 49

Practise:

4. 32 + 37
5. 28 + 51
6. 55 + 38
7. 21 + 39
8. 69 + 15
9. 44 + 38
10. 37 + 52

Answers

1. 16
2. 29
3. 18
4. 31
5. 19
6. 24
7. 31
8. 39
9. 71
10. 50

Starter activity 61

Resources
A board or flip chart.

Objective
Repeat addition or multiplication in a different order.

Strategies
• Write three additions on the board (see right). Ask for solutions.
• Discuss each in turn, emphasising that the order of addition has no effect on the answer.
• Encourage 'convenient' pairings, such as: numbers making a multiple of 10; doubles or near doubles; adding a 'near multiple of 10' and then adjusting.

Check it out

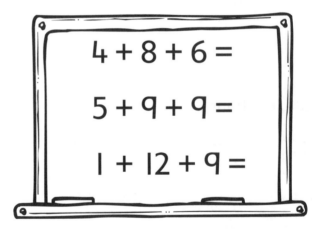

$$4 + 8 + 6 =$$

$$5 + 9 + 9 =$$

$$1 + 12 + 9 =$$

Write each of the questions below on the board and ask for the answer. Discuss some of them.

1. 6 + 3 + 7
2. 12 + 9 + 8
3. 5 + 4 + 9
4. 21 + 6 + 4
5. 7 + 7 + 5
6. 9 + 6 + 9
7. 12 + 13 + 6
8. 16 + 19 + 4
9. 31 + 30 + 10
10. 15 + 17 + 18

SCHOLASTIC

Near tens

Using the 100 square, take away 20 from...

1.   27

2.   43

3.   59

4.   30

5.   84

Take away 40 from...

6.   75

7.   42

8.   98

9.   61

10.   56

Take away 21 from...

11.   43

12.   88

13.   65

14.   39

15.   51

Take away 29 from...

16.   68

17.   93

18.   70

19.   59

20.   86

Starter activity 62

Resources
A 1–100 square (copy of photocopiable page 77) per child.

Objective
Subtract mentally a 'near multiple of 10' from a two-digit number... by subtracting 10, 20, 30... and adjusting.

Strategies
● Encourage making 'jumps' of 10.

● Ask 87 – 21 = ☐ to encourage the idea of taking away 20, then 1 more.

Answers

1.   7

2.   23

3.   39

4.   10

5.   64

6.   35

7.   2

8.   58

9.   21

10.   16

11.   22

12.   67

13.   44

14.   18

15.   30

16.   39

17.   64

18.   41

19.   30

20.   57

5× facts

Answers

Starter activity 63

Objective
Know by heart multiplication facts for the 5 times table.

Strategies
- Chant the 5 times table forwards and backwards.
- A rapid recall session.

1. 4×5
2. 10 multiplied by 5
3. What is the product of 6 and 5?
4. 1×5
5. 8 multiplied by 5
6. 5×5
7. double 5
8. What is the product of 9 and 5?
9. 7×5
10. 3×5
11. How many 5s in 10?
12. How many 5s make 30?
13. How many 5s make 5?
14. How many 5s make 35?
15. How many 5s make 50?
16. How many 5s make 20?
17. How many 5s make 45?
18. How many 5s make 15?
19. How many 5s make 40?
20. How many 5s make 25?

Starter activity 64

Objective
Say the number that is 1, 10, or 100 more or less than any given two- or three-digit number.

Clap counter: a game

Give the children a start number and an instruction (such as 'add 10').
Together they carry out the instruction and keep going until you clap. Then they change direction and carry on. For example:

22, 32, 42, 52, 62, clap, 52, 42, clap, 52, 62, 72... 'stop'

1. 28 add 10
2. 154 add 100
3. 102 add 1
4. 131 add 100
5. 492 add 1
6. 275 add 10
7. 87 take away 10
8. 333 take away 10
9. 206 take away 1
10. 987 take away 100

SCHOLASTIC

Place value

Ask pairs of children to repeat:

$7 \times 10 = 70$

$5 \times 10 = 50$

$14 \times 10 = 140$

$38 \times 10 = 380$

$26 \times 10 = 260$

Ask the following, with children raising their hands:

1. 25×10
2. 19×10
3. 12×10
4. 33×10
5. 41×10
6. 24×10
7. 37×10
8. 52×10
9. 76×10
10. 43×10

Starter activity 65

Resources
A board or flip chart; a set of numeral cards 0–9.

Objective
Shift the digits of a number one place to the left to multiply by 10.

Strategies
• Write the headings H, T and U on the board.
• Hold the '4' numeral card in the U column. Ask the children to multiply it by 10. Move the card into the T column and add a '0' card as a place holder.

Answers
1. 250
2. 190
3. 120
4. 330
5. 410
6. 240
7. 370
8. 520
9. 760
10. 430

Moving on up

Gather the children around you. Draw spaces for three two-digit numbers on the board:

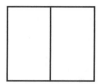  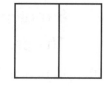

On each roll of the dice, the children decide in which space the number should be written. The aim is to end up with three two-digit numbers in ascending order.

Play a few times and gather strategies.

Starter activity 67

Resources
A board or flip chart, a dice.

Objective
Understand and use the vocabulary of comparing and ordering numbers.

3x facts

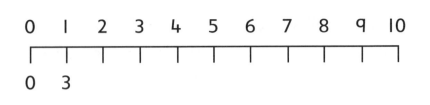

0 1 2 3 4 5 6 7 8 9 10

0   3

1. 6
2. 15
3. 30
4. 3
5. 12
6. 21
7. 3
8. 24
9. 18
10. 0
11. 27
12. 21
13. 9
14. 18
15. 8

Startor activity 66

Resources
A number line drawn on a board or flip chart; a pointer.

Objective
Begin to know the 3 times table.

Ask individual children to write the answers under the line. When the line is complete, count together in 3s forwards and backwards. Say the table together.

Point to numbers out of sequence, with children saying the complete number facts.

Remove all (or most) of the answers and ask quick-fire questions.

1. 2 times 3 equals
2. 5 multiplied by 3 equals
3. Ten 3s are
4. 1 × 3 =
5. 4 steps of 3 equal
6. 7 × 3 =
7. How many 3s make 9?
8. 8 multiplied by 3 equals
9. The product of 6 and 3 is
10. 0 times 3 equals
11. 9 × 3 =
12. 7 steps of 3 equal
13. How many 3s make 27?
14. 6 × 3 =
15. How many 3s make 24?

Moving on out

Round to the nearest 10:

1. 52
2. 29
3. 71
4. 34
5. 67
6. 45
7. 8
8. 13
9. 75
10. 86

Round to the nearest 100:

11. 178
12. 491
13. 234
14. 517
15. 350
16. 282
17. 303
18. 669
19. 450
20. 746

Starter activity 68

Answers

Resources
A 1–100 square (copy of photocopiable page 77) per child.

Objective
Round any two-digit number to the nearest 10 and any three-digit number to the nearest 100.

Strategies
● Ask: *What is 27 to the nearest ten? Is it nearer 20 or 30?* Explain that numbers ending in 5 are rounded up.

● Ask: *What is 183 to the nearest hundred? What is 250 to the nearest hundred?* Explain that numbers ending in 50 are rounded up.

1. 50
2. 30
3. 70
4. 30
5. 70
6. 50
7. 10
8. 10
9. 80
10. 90
11. 200
12. 500
13. 200
14. 500
15. 400
16. 300
17. 300
18. 700
19. 500
20. 700

Steps of 3

1. 3 × 3
2. 10 × 3
3. 5 × 3
4. 2 × 3
5. 7 × 3
6. 1 × 3
7. 9 × 3
8. 4 × 3
9. 8 × 3
10. 6 × 3

Count together in threes to no more than 50 from:
2, 7, 1, 8.

Starter activity 69

Answers

Objectives
Count on in steps of 3 from any small number to at least 50, then back again. Begin to know the 3 times table.

Strategies
● Count in threes from 0 to 30 and back to 0.
● Say together the 3 times table.
● Ask the children to raise their hands with answers.

1. 9
2. 30
3. 15
4. 6
5. 21
6. 3
7. 27
8. 12
9. 24
10. 18

SCHOLASTIC

55

Starter activity 70

Resources
A board or flip chart.

Objectives
Count on in steps of 4 from any small number to at least 50, then back again. Begin to know the 4 times table.

Steps of 4

Count together in fours from 0 to 40 and back again.

Write the 4 times table on the board as the children say it. Recite it together.

Divide the class into groups of four. The first group starts with '1 times 4 is 4'. The second group continues, and so on. Some of the answers can be removed as the count continues.

Repeat.

Answers

1. 20
2. 9
3. 40
4. 21
5. 4
6. 6
7. 30
8. 24
9. 12
10. 27
11. 5
12. 2
13. 8
14. 1
15. 6
16. 4
17. 9
18. 7
19. 3
20. 8

Starter activity 71

Objective
Begin to know the 3 and 4 times tables.

Strategies
● Recite the 3 and 4 times tables together.

3× and 4×

1. 5×4
2. 3×3
3. 10×4
4. 7×3
5. 1×4
6. 2×3
7. 10×3
8. 6×4
9. 4×3
10. 9×3

11. How many 3s make 15? How many 4s make 16?
12. How many 4s make 8? How many 4s make 36?
13. How many 4s make 32? How many 4s make 28?
14. How many 3s make 3? How many 4s make 12?
15. How many 3s make 18? How many 3s make 24?

Staging posts

How many more to 50 from...?

1. 28
2. 16
3. 20
4. 41
5. 33

6. 15
7. 22
8. 39
9. 27
10. 34

From 70, take away...

11. 42
12. 25
13. 38
14. 10
15. 24

16. 51
17. 36
18. 19
19. 43
20. 57

Starter activity 72

Resources
A 1–100 square (copy of photocopiable page 77) per child.

Objective
Extend understanding of the operations of addition and subtraction, and continue to recognise that addition can be done in any order.

Strategies
• Write 50 – 28 = . Ask for the answer.
• What is the children's preferred method? Encourage complementary addition.
• Stress that moving in 'jumps' of 10 makes counting up easier.

Answers

1. 22
2. 34
3. 30
4. 9
5. 17
6. 35
7. 28
8. 11
9. 23
10. 16

11. 28
12. 45
13. 32
14. 60
15. 46
16. 19
17. 34
18. 51
19. 27
20. 13

Staging posts

1. 50 – 28
2. 80 – 44
3. 100 – 79
4. 70 – 36
5. 60 – 42

6. 90 – 15
7. 100 – 61
8. 80 – 27
9. 40 – 13
10. 60 – 20

11. 100 – 52
12. 50 – 16
13. 40 – 28
14. 70 – 50
15. 90 – 61

16. 30 – 15
17. 80 – 37
18. 60 – 33
19. 90 – 49
20. 70 – 34

Starter activity 73

Resources
A 1–100 square (copy of photocopiable page 77) per child.

Objective
Extend understanding of the operation of addition and the related vocabulary, and continue to recognise that addition can be done in any order.

Strategies
• Encourage a variety of strategies. Ask some children to explain how they found the answer.
• Try questions 11–20 without 1–100 squares.

Answers

1. 22
2. 36
3. 21
4. 34
5. 18
6. 75
7. 39
8. 53
9. 27
10. 40

11. 48
12. 34
13. 12
14. 20
15. 29
16. 15
17. 43
18. 27
19. 41
20. 36

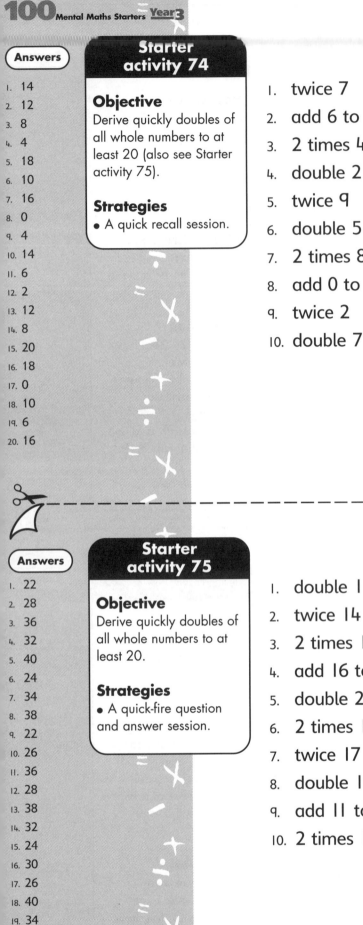

Double up

1. 14
2. 12
3. 8
4. 4
5. 18
6. 10
7. 16
8. 0
9. 4
10. 14
11. 6
12. 2
13. 12
14. 8
15. 20
16. 18
17. 0
18. 10
19. 6
20. 16

Starter activity 74

Objective
Derive quickly doubles of all whole numbers to at least 20 (also see Starter activity 75).

Strategies
• A quick recall session.

1. twice 7
2. add 6 to itself
3. 2 times 4
4. double 2
5. twice 9
6. double 5
7. 2 times 8
8. add 0 to 0
9. twice 2
10. double 7

11. 2 times 3
12. add 1 to itself
13. twice 6
14. add 4 to itself
15. double 10
16. add 9 to itself
17. double 0
18. twice 5
19. add 3 to 3
20. add 8 to itself

Double up again

1. 22
2. 28
3. 36
4. 32
5. 40
6. 24
7. 34
8. 38
9. 22
10. 26
11. 36
12. 28
13. 38
14. 32
15. 24
16. 30
17. 26
18. 40
19. 34
20. 30

Starter activity 75

Objective
Derive quickly doubles of all whole numbers to at least 20.

Strategies
• A quick-fire question and answer session.

1. double 11
2. twice 14
3. 2 times 18
4. add 16 to itself
5. double 20
6. 2 times 12
7. twice 17
8. double 19
9. add 11 to itself
10. 2 times 13

11. double 18
12. double 14
13. 2 times 19
14. twice 16
15. double 12
16. add 15 to itself
17. double 13
18. twice 20
19. 2 times 17
20. double 15

Near doubles

1. 6 + 7

2. 8 + 9

3. 13 + 12

4. 20 + 19

5. 15 + 16

6. 18 + 19

7. 22 + 21

8. 25 + 24

9. 16 + 17

10. 15 + 14

Starter activity 76

Answers

Resources
A board or flip chart.

Objective
Identify near-doubles, using doubles already known.

Strategies
● Write 14 + 15 = to show that the answer is equal to double 14 plus 1.
● Encourage compensating up or down as necessary.

---

11. 7 + 9

12. 10 + 12

13. 26 + 24

14. 22 + 20

15. 14 + 12

16. 17 + 19

17. 41 + 39

18. 16 + 14

19. 29 + 31

20. 11 + 13

● Write 21 + 19 = to show that a difference of 2 means the number in between can be doubled.

Answers

1. 13
2. 17
3. 25
4. 39
5. 31
6. 37
7. 43
8. 49
9. 33
10. 29

11. 16
12. 22
13. 50
14. 42
15. 26
16. 36
17. 80
18. 30
19. 60
20. 24

Near doubles

1. 45
2. 33
3. 23
4. 34
5. 61
6. 40

7. 141
8. 138
9. 159
10. 122
11. 103
12. 222
13. 181
14. 401
15. 602

Starter activity 77

Resources
A board or flip chart.

Objective
Identify near doubles, using doubles already known (eg 80 + 81).

Strategies
• Practise adding near doubles with totals below 100.

• Write 60 + 61 = . Ask for the answer and the method used.
• Discuss several examples, gathering a list of various strategies (eg double 6 tens = 12 tens = 120 and then 120 + 1 = 121).

1. 23 + 22
2. 17 + 16
3. 11 + 12

4. 16 + 18
5. 31 + 30
6. 21 + 19

7. 70 + 71
8. 70 + 68
9. 79 + 80
10. 60 + 62
11. 50 + 53

12. 110 + 112
13. 90 + 91
14. 200 + 201
15. 300 + 302

■ SCHOLASTIC

Pairs of 5s

Count in 5s to 100, forwards and backwards. Write the table of 5s to 50 as the children chant it together, then continue to 20 × 5 = 100. Erase the odd multiples of 5. Encourage strategies for working out answers beyond 10 × 5.

Ask individual children to give the answer and the complementary number that would take the answer to 100 (eg 12 × 5 = 60 → 40).

1. 8 × 5
2. 10 × 5
3. 14 × 5
4. 18 × 5
5. 16 × 5

6. 4 × 5
7. 12 × 5
8. 20 × 5
9. 2 × 5
10. 6 × 5

More pairs

As in activity 78, build a table of 5s extending to 20 × 5 = 100. Erase the even multiples. Suggest finding the complementary number taking an odd multiple up to 100 by adding 5, then counting up in tens to 100 (eg 11 × 5 = 55 → 55 + 5 = 60 → 60 + 40 = 100 → 40 + 5 = 45).

Ask individual children to give both the answer and the complementary number that would take the answer to 100.

1. 9 × 5
2. 7 × 5
3. 11 × 5
4. 5 × 5
5. 19 × 5

6. 13 × 5
7. 1 × 5
8. 15 × 5
9. 3 × 5
10. 17 × 5

I say, you say

The children together say the complementary number that makes the total up to 20.

Starter activity 80

Objective
Know by heart all addition and subtraction facts for each number to 20.

1. I say 12, you say...
2. I say 9, you say...
3. I say 5, you say...
4. I say 16, you say...
5. I say 10, you say...
6. I say 18, you say...
7. I say 15, you say...
8. I say 7, you say...
9. I say 11, you say...
10. I say 20, you say...
11. I say 14, you say...
12. I say 6, you say...
13. I say 13, you say...
14. I say 2, you say...
15. I say 17, you say...
16. I say 8, you say...
17. I say 4, you say...
18. I say 19, you say...
19. I say 3, you say...
20. I say 1, you say...

Place value

Write H T U on the board, with a 6 in the units column. Ask: *What number is 10 times larger than 6?* Ask an individual child to write the answer on the board. Emphasise that the 6 has moved across the 'boundary' from units to tens, and that the 0 acts as a 'place holder'. Ask: *What number is 10 times larger than 60?* Repeat the above.

Write each starting number and ask for a volunteer to multiply it by 10.

1.	9	6.	5
2.	90	7.	30
3.	2	8.	7
4.	20	9.	10
5.	8	10.	4

Starter activity 81

Resources
A board or flip chart.

Objective
Shift the digits one place to the left to multiply by 10.

Answers

1. 90
2. 900
3. 20
4. 200
5. 80
6. 50
7. 300
8. 70
9. 100
10. 40

Place value

As in Starter activity 81, prepare the board with a 6 in the units column.

Ask: *What number is 100 times larger than 6?* Emphasise that the 6 has crossed the two 'boundaries' from units to hundreds, and that two zeros have been added as 'place holders'.

Write each starting number and ask for a volunteer to multiply it by 100.

1.	4	5.	8
2.	7	6.	5
3.	6	7.	1
4.	2	8.	9

Starter activity 82

Resources
A board or flip chart.

Objective
Shift the digits two places to the left to multiply by 100.

Answers

1. 400
2. 700
3. 600
4. 200
5. 800
6. 500
7. 100
8. 900

Twice as much

1. 20
2. 14
3. 36
4. 28
5. 18

6. 17
7. 20
8. 30
9. 29
10. 20

11. 10
12. 4
13. 17
14. 6
15. 2

16. 17
17. 3
18. 22
19. 20
20. 27

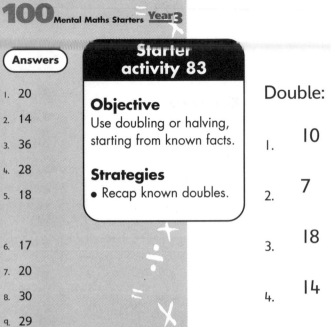

Starter activity 83

Objective
Use doubling or halving, starting from known facts.

Strategies
● Recap known doubles.

● Recap known halves.

Double:

1. 10

2. 7

3. 18

4. 14

5. 9

Half of:

6. 20

7. 8

8. 34

9. 12

10. 4

Ask:

11. double 6 plus 5

12. double 13 minus 6

13. double 5 times 3

14. double 16 minus 3

15. double 8 plus 4

Ask:

16. half of 22, plus 6

17. half of 16, minus 5

18. half of 38, plus 3

19. half of 10, times 4

20. half of 18, times 3

÷2 and ÷10

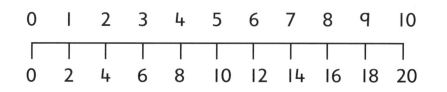

1. How many 2s make 10?
2. How many 2s make 4?
3. Divide 12 by 2.
4. Divide 8 by 2.
5. 18 divided by 2 is...
6. 2 divided by 2 is...
7. How many 2s make 14?
8. Divide 20 by 2.
9. 6 divided by 2 is...
10. How many 2s make 16?

Repeat with the number line erased.

Recall the division facts for the 10 times table.

11. How many 10s make 50?
12. How many 10s make 70?
13. 20 divided by 10 equals...
14. 100 divided by 10 equals...
15. Divide 40 by 10.
16. Divide 60 by 10.
17. How many 10s make 90?
18. 10 divided by 10 equals...
19. How many 10s make 80?
20. Divide 30 by 10.

Starter activity 84

Resources
A board or flip chart.

Objectives
Recognise that division is the inverse of multiplication.
Derive division facts corresponding to the 2 and 10 times tables.

Strategies
• Draw a number line for the 2 times table, with answers, on the board.
• Recite the 2 times table together, saying each complete fact ('1 times 2 is 2', etc). Recite the table as a division table ('2 divided by 2 is 1', etc). Emphasise that division is the inverse of multiplication.

Answers

1. 5
2. 2
3. 6
4. 4
5. 9
6. 1
7. 7
8. 10
9. 3
10 8
11. 5
12. 7
13. 2
14. 10
15. 4
16. 6
17. 9
18. 1
19. 8
20. 3

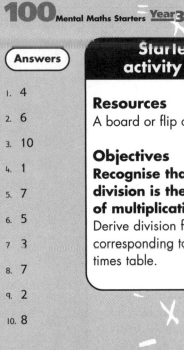

Answers

1. 4
2. 6
3. 10
4. 1
5. 7
6. 5
7. 3
8. 7
9. 2
10. 8

Resources
A board or flip chart.

Objectives
Recognise that division is the inverse of multiplication.
Derive division facts corresponding to the 5 times table.

÷5

Draw a number line for the 5 times table, with answers, on the board.

```
0   1   2   3   4   5   6   7   8   9   10
|   |   |   |   |   |   |   |   |   |   |
0   5  10  15  20  25  30  35  40  45  50
```

Recite the 5 times table together. Recite the table together as a division table.

1. How many 5s make 20?
2. Divide 30 by 5.
3. 50 divided by 5 is...
4. Divide 5 by 5.
5. How many 5s in 35?
6. How many 5s in 25?
7. 15 divided by 5 is...
8. 35 divided by 5 is...
9. How many 5s in 10?
10. Divide 40 by 5.

Repeat with the number line erased.

Half of half

Write 32 → 16 → 8 to demonstrate finding a quarter by halving twice: 'half and half again'.

Find half and half again of:

1.	16	7.	24
2.	40	8.	36
3.	8	9.	48
4.	20	10.	32
5.	12	11.	44
6.	4	12.	28

Conclude that to find a quarter, we find 'half and half again' or divide by 4. If time allows, repeat the questions, asking for 'a quarter of...'

Starter activity 86

Objective
Use doubling or halving, starting from known facts.

Answers
1. 8, 4
2. 20, 10
3. 4, 2
4. 10, 5
5. 6, 3
6. 2, 1
7. 12, 6
8. 18, 9
9. 24, 12
10. 16, 8
11. 22, 11
12. 14, 7

- -

Half of half

Find half and half again of:

1.	100	6.	60
2.	200	7.	800
3.	88	8.	120
4.	400	9.	240
5.	80	10.	160

Starter activity 87

Objective
Use doubling or halving, starting from known facts.

Answers
1. 50, 25
2. 100, 50
3. 44, 22
4. 200, 100
5. 40, 20
6. 30, 15
7. 400, 200
8. 60, 30
9. 120, 60
10. 80, 40

Shift the digit

Starter activity 88

Resources
A board or flip chart with prepared HTU columns.

Objective
Shift the digits one/two places to the left to multiply by 10/100.

Strategies
• Write each number in turn on the HTU chart and ask individual children to write the answers in the correct columns beneath.
• Remind the class that zeros will be needed to hold the digits in place.

1. 50
2. 400
3. 900
4. 700
5. 20
6. 300
7. 800
8. 300
9. 80
10. 600

1. 5, make 10 times larger
2. 40, make 10 times larger
3. 9, make 100 times larger
4. 7, make 100 times larger
5. 2, make 10 times larger
6. 30, make 10 times larger
7. 80, make 10 times larger
8. 3, make 100 times larger
9. 8, make 10 times larger
10. 60, make 10 times larger

Shift the digit 2

Starter activity 89

Resources
A board or flip chart with prepared HTU columns.

Objective
Recognise that division is the inverse of multiplication.

Strategies
• Write 50. Ask a child to make it 10 times smaller, writing the answer in the correct columns.
• Write 300 and ask for it to be made a hundred times smaller.
• Repeat for the examples, asking a child to write each answer in the correct columns.

1. 7
2. 2
3. 4
4. 6
5. 90
6. 5
7. 3
8. 2
9. 80
10. 6

1. 70, make 10 times smaller
2. 200, make 100 times smaller
3. 40, make 10 times smaller
4. 60, make 10 times smaller
5. 900, make 10 times smaller
6. 500, make 100 times smaller
7. 30, make 10 times smaller
8. 20, make 10 times smaller
9. 800, make 10 times smaller
10. 600, make 100 times smaller

■SCHOLASTIC

100 square facts

Ask the questions. The children shade the answers on the 1–100 square.

1. 5 plus 7
2. 15 minus 9
3. half of 36
4. double 12
5. 20 + 10

6. 2 + 8 + 6 + 4 + 1
7. the product of 3 and 5 is
8. 26 minus 17
9. 21 minus 18
10. double 13 plus 1

Ask: *Which times table are we shading?*

The children identify and shade the remaining squares in the times table pattern.

Starter activity 90

Resources
A pencil and a 1–100 square (copy of photocopiable page 77) per child.

Objective
Begin to know the 3 times table.

Answers
1. 12
2. 6
3. 18
4. 24
5. 30
6. 21
7. 15
8. 9
9. 3
10. 27

×3 and ÷3

1. 5×3
2. How many 3s in 15?
3. How many steps of 3 to 30?
4. 2×3
5. 8 multiplied by 3
6. divide 24 by 3
7. 9 divided by 3
8. 1×3
9. multiply 3 by 9
10. 3×3

11. How many 3s make 12?
12. How many 3s make 21?
13. divide 6 by 3
14. 4 multiplied by 3
15. 3 divided by 3
16. 18 divided by 3
17. 10 times 3
18. 7 times 3
19. How many 3s in 27?
20. 6 multiplied by 3

Starter activity 91

Resources
A board or flip chart.

Objective
Explain methods and reasoning orally and, where appropriate, in writing.

Strategies
● Write $7 \times 3 = 21$ on the board and ask for a division fact from it.

Answers
1. 15
2. 5
3. 10
4. 6
5. 24
6. 8
7. 3
8. 3
9. 27
10. 9
11. 4
12. 7
13. 2
14. 12
15. 1
16. 6
17. 30
18. 21
19. 9
20. 18

×4 and ÷4

1. 12
2. 3
3. 28
4. 7
5. 8
6. 36
7. 4
8. 8
9. 24
10. 10
11. 1
12. 20
13. 40
14. 32
15. 6
16. 2
17. 9
18. 5
19. 16
20. 4

Starter activity 92

Objectives
Begin to know the 4 times table. **Recognise that division is the inverse of multiplication.**

Strategies
Ask the children for a division fact from 5 × 4 = 20. Use this to remind them of the connection between multiplication and division.

1. 3 × 4
2. How many 4s in 12?
3. 7 × 4
4. divide 28 by 4
5. 2 × 4
6. 9 × 4
7. How many 4s make 16?
8. How many 4s make 32?
9. multiply 6 by 4
10. divide 40 by 4
11. How many 4s make 4?
12. 5 × 4
13. 10 × 4
14. 8 multiplied by 4
15. share 24 between 4
16. divide 8 by 4
17. How many 4s make 36?
18. How many 4s make 20?
19. 4 × 4
20. multiply 4 by 1

Target totals

1. 9 + 7 + 4
2. 9 + 7 + 3
3. 9 + 7 **or** 9 + 3 + 4
4. 9 + 7 + 3 + 4
5. 11 + 6 **or** 11 + 4 + 2
6. 6 + 4 + 2
7. 11 + 6 + 4
8. 11 + 6 + 4 + 2

Starter activity 93

Resources
A board or flip chart; pencils and paper.

Objective
Understand that more than two numbers can be added together.

Strategies
● Divide the class into mixed-ability pairs.
● Use the first example as a reminder of addition skills previously taught (eg putting the larger number first, making multiples of 10, partitioning and recombining).

Write the numbers 3, 4, 7 and 9. Ask the children to make 14 by adding two or more numbers. Set more target totals:

1. 20
2. 19
3. 16
4. 23

Now write the numbers 2, 4, 6 and 11. Ask the children to make:

5. 17
6. 12
7. 21
8. 23

SCHOLASTIC

Target products

Write the numbers 2, 3, 4 and 5. Ask the children to multiply two or more of these numbers to 'hit' the target totals.

1. 10
2. 20
3. 40
4. 15
5. 30

6. 24
7. 60
8. 120
9. 6
10. 12

Answers

1. 2 × 5
2. 4 × 5
3. 5 × 2 × 4
4. 3 × 5
5. 5 × 3 × 2
6. 3 × 4 × 2
7. 4 × 5 × 3
8. 3 × 4 × 2 × 5
9. 2 × 3
10. 3 × 4

Any time

Move the minute hand from the 'o'clock' position, together saying the time every 5 minutes (eg '5 minutes past...' '25 minutes to...') all the way round.

The children show and say these times on their clock faces.

1. 10 o'clock
2. 10 minutes past 10
3. 20 minutes past 10
4. 5 minutes past 10
5. 15 minutes past 10
6. 25 minutes past 10

7. 30 minutes past 10
8. 5 minutes to 11
9. 25 minutes to 11
10. 10 minutes to 11
11. 20 minutes to 11
12. 15 minutes to 11

Repeat, starting at 6 o'clock.

Starter activity 95

Resources
A teaching clock.

Objectives
Understand and use the vocabulary related to time. Read the time to 5 minutes on an analogue clock.

Take five

Count in 5s to 60 and back again.

Move the minute hand round the clock face as the children say together '5 minutes past', '10 minutes past'... up to '30 minutes past' (or 'half past'). Move the minute hand to specific times, the children together saying:

1. 10 minutes past
2. 20 minutes past
3. 5 minutes past
4. 25 minutes past
5. 15 minutes past (or quarter past)
6. 30 minutes past (or half past)

Repeat.

Move the minute hand from half past onwards, the children saying together '25 minutes to', '20 minutes to'... up to the hour. Move the minute hand to specific times, the children together saying:

7. 20 minutes to
8. 5 minutes to
9. 25 minutes to
10 15 minutes to (or quarter to)
11. 10 minutes to
12. o'clock

Repeat.

Select times at random around the clock to test the children.

▲ S C H O L A S T I C

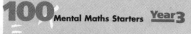

5 minutes later

Ask the children to take turns (within each pair) to show the times that you say.

Starter activity 97

Resources
A teaching clock; one clock face (as in activity 96) between two children.

Objective
Understand and use the vocabulary related to time.

1.   10 minutes past 9
2.   20 minutes past 12
3.   25 minutes past 7
4.   5 minutes past 11
5.   quarter past 6
6.   half past 8
7.   5 minutes to 4
8.   25 minutes to 1
9.   20 minutes to 5
10.   10 minutes to 3

Repeat, asking the children to advance each time by five minutes and say the new time.

Ask the class:
- *How many minutes in one hour?*
- *How many minutes in half an hour?*
- *How many minutes in quarter of an hour?*
- *How many minutes in three quarters of an hour?*

Tens

Starter activity 98

Resources
A board or flip chart.

Objective
Use known number facts and place value to multiply mentally.

Strategies
● Write 30 × 5 = . Ask for the answer and the strategies used.
● Stress that when we multiply by 10 (for example, 30 × 5 = 5 × 3 × 10 = 15 × 10 = 150), a zero is needed as a place holder.
● Discuss several of these examples.

Answers

1. 60
2. 400
3. 200
4. 60
5. 120
6. 200
7. 150
8. 300
9. 400
10. 160
11. 240
12. 320
13. 800
14. 180
15. 250

1. 6 × 10
2. 40 × 10
3. 20 × 10
4. 20 × 3
5. 40 × 3
6. 40 × 5
7. 30 × 5
8. 60 × 5
9. 80 × 5
10. 40 × 4
11. 60 × 4
12. 80 × 4
13. 80 × 10
14. 60 × 3
15. 50 × 5

Tens

Starter activity 100

Objective
Use knowledge of number facts and place value to multiply and divide mentally.

Strategies
● Discuss several of the questions and the strategies that have been used. Stress that when we make the number bigger by multiplying, the digits move to the left. When we make a number smaller by dividing, the digits move to the right.

Answers

1. 50
2. 80
3. 200
4. 600
5. 7
6. 9
7. 1
8. 3
9. 200
10. 350
11. 300
12. 10
13. 100
14. 10
15. 60
16. 300
17. 800
18. 240
19. 200
20. 180

1. 5 × 10
2. 8 × 10
3. 20 × 10
4. 60 × 10
5. 70 ÷ 10
6. 90 ÷ 10
7. 10 ÷ 10
8. 30 ÷ 10
9. 40 × 5
10. 70 × 5
11. 60 × 5
12. 80 ÷ ☐ = 8
13. 900 ÷ ☐ = 9
14. 500 ÷ ☐ = 50
15. ☐ ÷ 10 = 6
16. ☐ ÷ 100 = 3
17. ☐ ÷ 10 = 80
18. 80 × 3
19. 50 × 4
20. 60 × 3

Tens

1. $70 \div 10 =$

2. $40 \div 10 =$

3. $800 \div 10 =$

4. $500 \div 10 =$

5. $600 \div 100 =$

6. $300 \div 100 =$

7. $900 \div \square = 90$

8. $50 \div \square = 5$

9. $700 \div \square = 7$

10. $500 \div \square = 5$

11. $\square \div 10 = 40$

12. $\square \div 100 = 8$

13. $\square \div 10 = 30$

14. $\square \div 100 = 2$

15. $\square \div 10 = 5$

Starter activity 99

Resources
A board or flip chart.

Objectives
Recognise that division is the inverse of multiplication. Use knowledge of number facts and place value to divide mentally.

Strategies
• Write $60 \div 10 =$. Ask for the answer and the strategies used.
• Stress that when we shift digits to the right in order to divide by 10 or 100, the zero place holder(s) must be removed.

• Write these statements on the board and ask: *What have we divided by?*

• Ask: *Can these unknown numbers be found?*

Answers

1. 7
2. 4
3. 80
4. 50
5. 6
6. 3

7. 10
8. 10
9. 100
10. 100

11. 400
12. 800
13. 300
14. 200
15. 50

Hundreds, tens and units chart

1	2	3	4	5	6	7	8	9
100	200	300	400	500	600	700	800	900
10	20	30	40	50	60	70	80	90

Enlarge to at least A3 size for whole-class use.

SCHOLASTIC

1–100 square

1	2	3	4	5	6	7	8	9	10
11	12	13	14	15	16	17	18	19	20
21	22	23	24	25	26	27	28	29	30
31	32	33	34	35	36	37	38	39	40
41	42	43	44	45	46	47	48	49	50
51	52	53	54	55	56	57	58	59	60
61	62	63	64	65	66	67	68	69	70
71	72	73	74	75	76	77	78	79	80
81	82	83	84	85	86	87	88	89	90
91	92	93	94	95	96	97	98	99	100

Support for Starter activity 47

Choose a colour.

Colour 2 quarters of each shape and write in the spaces.

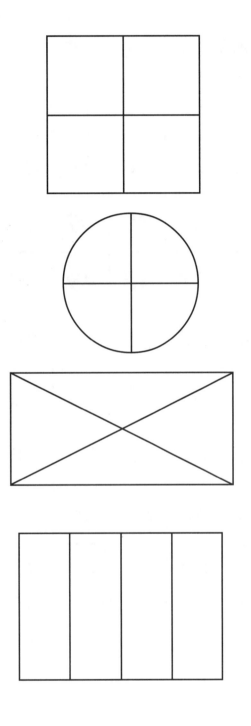

I have coloured two quarters _____.

I have coloured one half _____.

$$\frac{1}{4} \ + \ \frac{1}{4} \ = \ \frac{1}{2}$$

I have coloured two _____ _____.

I have coloured one _____ _____.

$$\frac{1}{4} \ + \ \frac{1}{4} \ =$$

I have coloured two _____ _____.

I have coloured one _____ _____.

$$\frac{1}{4} \ + \quad \ =$$

I have coloured ____ _____ _____.

I have coloured ____ _____ _____.

$$+ \quad =$$

Support for Starter activity 49

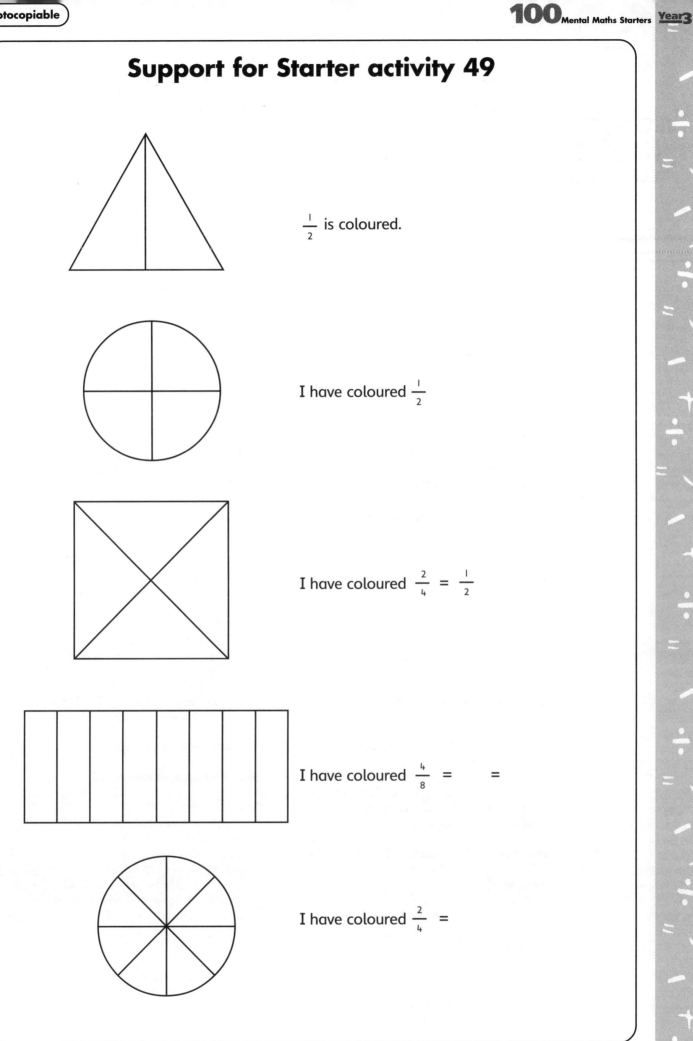

$\frac{1}{2}$ is coloured.

I have coloured $\frac{1}{2}$

I have coloured $\frac{2}{4}$ = $\frac{1}{2}$

I have coloured $\frac{4}{8}$ =   =

I have coloured $\frac{2}{4}$ =

Index

Note that the numbers given here are activity numbers, not page numbers.